MERCHANT SHIP STABILITY

MERCHANT SHIP STABILITY

(METRIC EDITION)

A Companion to "Merchant Ship Construction"

BY

H. J. PURSEY

EXTRA MASTER

Formerly Lecturer to the School of Navigation

University of Southampton

GLASGOW

BROWN, SON & FERGUSON, LTD., NAUTICAL PUBLISHERS

4-10 DARNLEY STREET

First Edition	–	1945
Sixth Edition	–	1977
Revised	–	1983
Reprinted	–	1992
Reprinted	–	1996

ISBN 0 85174 442 7 (Revised Sixth Edition)
ISBN 0 85174 274 2 (Sixth Edition)

Printed and Made in Great Britain

INTRODUCTION

DURING the past few years there have been considerable changes in the approach to ship stability, so far as it affects the merchant seaman. The most obvious of these is the introduction of metric units. In addition, the Department of Trade have already increased their examination requirements: they have also produced recommendations for a standard method of presenting and using stability information, which will undoubtedly be reflected in the various examinations.

This revised edition has been designed to meet the above-mentioned requirements. The basic information contained in the early chapters has been retained for the benefit of those who are not familiar with such matters. The remainder of the text has been re-arranged and expanded, as desirable, to lead into the new material which has been introduced; whilst a new chapter on stability information has been added to illustrate the Department of Trade recommendations.

The theory of stability has been covered up to the standard required for a Master's Certificate and includes all that is needed by students for Ordinary National Diplomas and similar courses. This has been carefully linked-up with practice, since the connection between the two is a common stumbling block. Particular attention has been paid to matters which are commonly misunderstood, or not fully appreciated by seamen.

H. J. P.

SOUTHAMPTON, 1982.

CONTENTS

CHAPTER 1—SOME GENERAL INFORMATION

CHAPTER 2—AREAS AND VOLUMES

CHAPTER 3—FORCES AND MOMENTS

CHAPTER 4—DENSITY, DEADWEIGHT AND DRAFT

CHAPTER 5—CENTRE OF GRAVITY OF SHIPS

CHAPTER 6—CENTRES OF BUOYANCY AND FLOTATION

CHAPTER 7—THE RIGHTING LEVER AND METACENTRE

CHAPTER 8—TRANSVERSE STATICAL STABILITY

CHAPTER 9—FREE SURFACE EFFECT

CHAPTER 10—TRANSVERSE STATICAL STABILITY IN PRACTICE

CHAPTER 11—DYNAMICAL STABILITY

CHAPTER 12—LONGITUDINAL STABILITY

CHAPTER 13—STABILITY CURVES AND SCALES

CHAPTER 14—BILGING OF COMPARTMENTS

CHAPTER 15—STABILITY AND THE LOAD LINE RULES

CHAPTER 16—MISCELLANEOUS MATTERS

CHAPTER 17—ROLLING

CHAPTER 18—SUMMARY

MERCHANT SHIP STABILITY

CHAPTER 1

THE METRIC SYSTEM

Length.—The basic unit of length is the Metre. 1 metre (m) = 10 decimetres (dm) = 100 centimetres (cm) = 1000 millimetres (mm).

Weight.—The metric ton, or tonne, is the weight of 1 cubic metre of fresh water. 1 tonne = 1000 kilogrammes (kg) = 1,000,000 grammes (gr). The gramme is the weight of 1 cubic centimetre of fresh water.

Volume.—Is measured in cubic metres (m^{-3}), or cubic centimetres (cc, or cm^{-3}).

Area.—Is measured in square metres (m^{-2}), or square centimetres (cm^{-2}).

Force.—When a force is exerted, it is usually measured, in stability, in tonnes or kilogrammes. To indicate that it is a force or weight, as distinct from mass, an 'f' is often added ; e.g. "tonnes f", or "kilogrammes f."

Moment.—For our purpose, this is usually expressed as tonne-metres (tonne f-m).

Pressure.—May be given as tonnes per square metre (tonnes f/m^2), or as kilogrammes per square centimetre (kg f/cm^2).

Density.—This is usually defined as mass per unit volume. For our purpose it can be regarded as the weight of one cubic metre or of one cubic centimetre of a substance. We may express it as either:

Grammes per cubic centimetre (grs/cm^3).

Kilogrammes per cubic metre (kg/m^3).

Tonnes per cubic metre ($tonnes/m^3$).

Relative Density.—Was formerly called "specific gravity". It is the ratio between the density of a substance and the density of fresh water.

$$\text{i.e. Relative density} = \frac{\text{density of substance}}{\text{density of fresh water}}$$

For simplicity the "f" and the negative indices are omitted hereafter: e.g. moment is written as tonne-metres, whilst, say, cubic metres are given as m^3.

Water.—The following values are usually taken for purposes of stability :—

	Fresh water	*Salt water*
Density (grs/cm^3 or tonnes/m^3)	1·000	1·025
Density (kg/m^3)	1000	1025
Relative density	1·000	1·025
Weight per cubic metre	1·000	1·025 tonnes
Weight per cubic metre	1000	1025 kilogrammes
Volume per tonne (cubic metres)	1·000	$\frac{1}{1\cdot025}$

Increase of Pressure with Depth.—The pressure on an object which is placed under water is equal to the column of water above it.

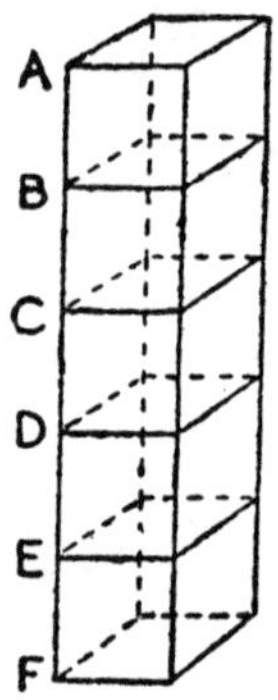

FIG. 1

Consider Fig. 1, which represents a column of water having an area of one square metre. Let A, B, C, D, E and F be points one metre apart vertically. The volume of water above B is one cubic metre; above C, two cubic metres; above D, three cubic metres; and so on. If δ is the density of the water in tonnes per cubic metre, the weight above B will be δ tonnes; above C will be 2δ tonnes; above D will be 3δ tonnes; and so on. We can see from this that if point A is at the sea surface, then the pressure per square metre at a depth of, say, AF metres, will be $AF \times \delta$ tonnes.

From the above, it is obvious that the pressure at any depth, in tonnes per square metre, is equal to δ times the depth in metres. Since water exerts pressure equally in all directions, this pressure will be the same horizontally, vertically, or obliquely. We can say, then, that if a horizontal surface of area A square metres is placed at a depth of D metres below the surface, then:—

$$\text{Pressure per square metre} = \delta \times D \text{ tonnes}$$
$$\text{Total pressure on the area} = \delta \times AD \text{ tonnes}$$

The Effect of Water in Sounding Pipes, etc.—When water rises in sounding pipes or air pipes to above the top of a tank, pressure is set up on the tank-top. The actual weight of water in the pipe may be comparatively small, but its effect may be considerable. Water exerts pressure equally in all directions and so the pressure per square centimetre at the bottom of the pipe is transmitted over the whole of the tank-top. This pressure will not depend on the actual weight of water in the pipe, but on the head of water and will be approximately the same whatever the diameter of the pipe. For this reason, tanks should not be left "pressed up" for long periods, because this can exert considerable stress on the tank-top.

Example.—A rectangular double bottom tank is being filled with sea water. If the water is allowed to rise in the sounding pipe to a height of 7 metres above the tank top, find the pressure per square metre on the tank top.

$$\text{Pressure} = \delta D = 1{\cdot}025 \times 7 = 7{\cdot}175\ \text{t/m}^2$$

The Law of Archimedes.—A body immersed in a liquid appears to suffer a loss in weight equal to the weight of liquid which it displaces. From this, we conclude that a floating body displaces its own weight of water. This can be shewn as follows:—

A block of iron, one cubic metre in size and of density 8·000 tonnes/m³ would weigh 8 tonnes in air. If placed in fresh water it would displace one cubic metre of water, which would weigh one tonne; so the weight of the block would thus appear to be 7 tonnes when it was under water.

If we now take the block and make it into a hollow, sealed box, its weight in air will remain the same but its volume will increase. If placed in water, it would displace more of the water and its apparent weight will decrease accordingly. For instance, if the box were 3 cubic metres in volume it would displace 3 cubic metres of water (or 3 tonnes), so that its apparent weight in fresh water would now be 5 tonnes.

If we increase the volume of the box still further, it will displace still more water and its apparent weight under water will decrease still more.

Eventually, when the volume of the box became greater than 8 cubic metres, an equivalent volume of water would weigh more than the box. So if the box were now placed under water, it would be forced upwards, until the upward force exactly equalled the weight of the box. In other words, the box would rise until it floated at such a draft that it would displace its own weight of water.

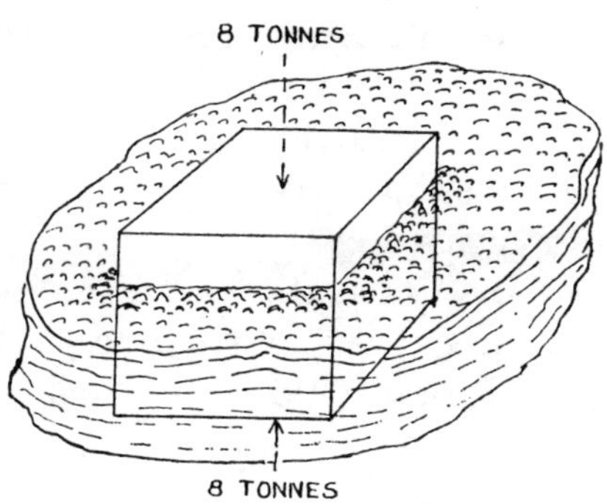

FIG. 2.

Application to Ships.—A ship may be regarded as a closed iron box, so that two conclusions can be drawn from a study of the last section:—

(*a*) So long as the weight of the ship does not exceed the weight of its own volume of water, it will float.

(*b*) The draft at which it floats will be such that the weight of water displaced will be equal to the weight of the ship.

Floating Bodies and the Density of Water.—Weight is equal to volume multiplied by density and therefore volume is equal to weight divided by density. In the case of a floating body:—

$$\text{Volume of water displaced} = \frac{\text{Weight of water displaced}}{\text{Density of the water}}$$

The volume of water displaced is equal to the underwater volume of the body and since a floating body displaces its own weight of water, weight of the body can be substituted for weight of water displaced.

$$\text{Underwater volume} = \frac{\text{Weight of the body}}{\text{Density of the water}}$$

If the weight of the body remains the same, it can be seen that the underwater volume must vary inversely as the density of the water. In other words, if the density of the water changes:—

$$\frac{\text{New volume displaced}}{\text{Old volume displaced}} = \frac{\text{Old density}}{\text{New density}}$$

Ship Dimensions.—The following are the principal dimensions used in measuring ships.

Lloyds' Length is the length of the ship, measured from the fore side of the stem to the after side of the stern post at the summer load-line. In ships with cruiser sterns, it is taken as 96 per cent of the length overall provided that this is not less than the above.

Moulded Breadth is the greatest breadth of the ship, measured from side to side outside the frames, but inside the shell plating.

Moulded Depth is measured vertically at the middle length of the ship, from the top of the keel to the top of the beams at the side of the uppermost continuous deck.

The Framing Depth is measured vertically from the top of the double bottom to the top of the beams at the side of the lowest deck.

Depth of Hold is measured at the centre line, from the top of the beams at the tonnage deck to the top of the double bottom or ceiling.

Decks.—The Freeboard Deck is the uppermost complete deck, having permanent means of closing all openings in its weather portion.

The Tonnage Deck is the upper deck in single-decked ships and the second deck in all others.

Ship Tonnages.—These are not measures of weight, but of space: the word "ton" being used to indicate 100 cubic feet or 2·83 cubic metres. For instance, if the gross tonnage of a ship is 5000 tons this does not mean that she weighs that amount, but that certain spaces in her measure 500,000 cubic feet or 14150 cubic metres.

Under Deck Tonnage is the volume of the ship below the tonnage deck. It does not normally include the cellular double bottom below the inner bottom: or, in the case of open floors, the space between the outer bottom and the tops of the floors.

Gross Tonnage is under deck tonnage, plus spaces in the hull above the tonnage deck. It also includes permanently enclosed superstructures, with some exceptions, and any deck cargo that is on board.

Nett Tonnage is found by deducting, from the gross tonnage, certain non-earning spaces. These "deductions" include crew accommodation, stores and certain water ballast spaces: also an "allowance for propelling power" which depends partly on the size of the machinery spaces.

Under the 1967 Tonnage Rules, some ships may now have a Modified Tonnage. This means that they have a tonnage which is less than the normal tonnage for a ship of their size, but are not allowed to load so deeply. Other ships may have two Alternative Tonnages: normal tonnage for use when they are loaded to their normal loadlines; or a modified tonnage when they are loaded less deeply. Such ships are marked with a special "Tonnage Mark" to indicate which tonnage is to be used.

Grain and Bale Measurement.—These terms are often found on the plans of ships and refer to the volume of the holds, etc.

Grain Measurement is the space in a compartment taken right out to the ship's side. In other words, it is the amount of space which would be available for a bulk cargo such as grain.

Bale Measurement is the space in a compartment measured to the inside of the spar ceiling, or, if this is not fitted, to the inside of the frames. It is the space which would be available for bales and similar cargoes.

Displacement.—Is the actual weight of the ship and all aboard her at any particular time. Since a floating body displaces its own weight of water, this means that displacement is equal to the weight of water displaced by the ship.

Light Displacement is that of the ship when she is at her designed light draft. It consists of the weight of the hull, machinery, spare parts and water in the boilers.

Loaded Displacement is that of a ship when she is floating at her summer draft.

Deadweight.—This is the weight of cargo, stores, bunkers, etc., on board a ship. In other words, it is the difference between the light displacement and the displacement at any particular draft. When we say that a ship is of so many tonnes deadweight, we usually mean that the difference between her light and loaded displacements is so many tonnes.

Draft.—This is the depth of the bottom of the ship's keel below the surface of the water. It is measured forward and aft at the ends of the ship. When the drafts at each end are the same, the ship is said to be on an even keel. When they differ, the ship is said to be trimmed by the head, or by the stern, according to which is the greater of the two drafts.

Mean Draft is the mean of the drafts forward and aft.

Freeboard.—Statutory Freeboard is the distance from the deck-line to the centre of the plimsoll mark. The term "Freeboard" is often taken to mean the distance from the deck-line to the water.

Ordinary Load-lines.—The load-lines and deck line must be painted in white or yellow on a dark background, or in black on a light background.

The deck-line is placed amidships and is 300 millimetres long and 25 millimetres wide. Its upper edge marks the level at which the top of the freeboard deck, if continued outward, would cut the outside of the shell plating.

A load-line disc, commonly called "the plimsoll mark", is placed below the deck-line. The distance from the upper edge of the deck-line to the centre of the disc is the statutory summer freeboard. 540 millimetres forward of the disc are placed the load-lines, which mark the drafts to which the ship may be loaded when at sea and in certain zones. All lines are 25 millimetres thick and their upper edges mark the level to which they refer. The following are the marks required for steam-ships:—

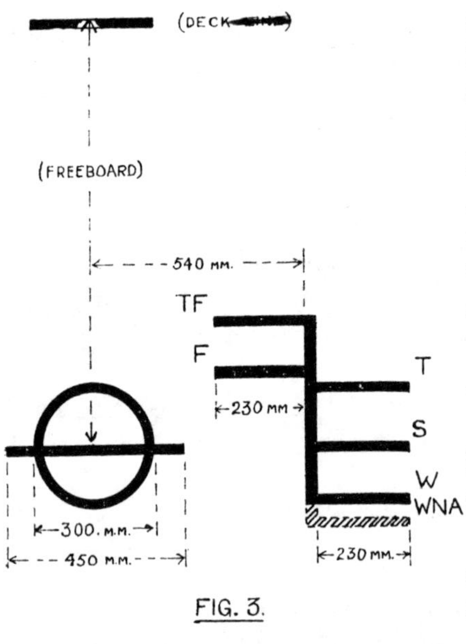

FIG. 3.

S—the "summer load-line"—is level with the centre of the disc.

W—the "winter load-line"—is placed below the summer load-line at a distance of one forty-eighth of summer draft.

T—the "tropical load-line"—is placed above the summer load-line at a distance of one forty-eighth of summer draft.

WNA—the "winter North Atlantic load-line"—is placed 50 millimetres below the winter load-line. It is only marked on ships which are 100 metres or less in length.

F—The "fresh-water (summer) load-line"—indicates the draft to which the ship can be loaded in fresh water, if she is to rise to her summer load-line at sea. Its distance above the summer load-line (*S*) is called the "Fresh Water Allowance" and is found by the formula $\frac{\Delta}{4T}$ millimetres, where Δ equals the displacement at summer draft and *T* is the "tonnes per centimetre immersion" at that draft.

TF—The "tropical fresh water load-line"—is the fresh-water line to which the ship can be loaded, in order that she may come to her tropical mark when she reaches the sea. It is found as for *F*, but is measured above the tropical line.

Timber Load-lines.—These are marked-in abaft the load-line disc. They show the drafts to which a ship so marked may load when carrying a deck cargo of timber, provided that the cargo is stowed according to special rules and to a certain minimum height.

LS—The "summer timber load-line"—This is a little above the ordinary summer line. The distance is such that if a ship, loaded according to the regulations, lost her deck cargo, she would come back approximately to her ordinary summer load-line.

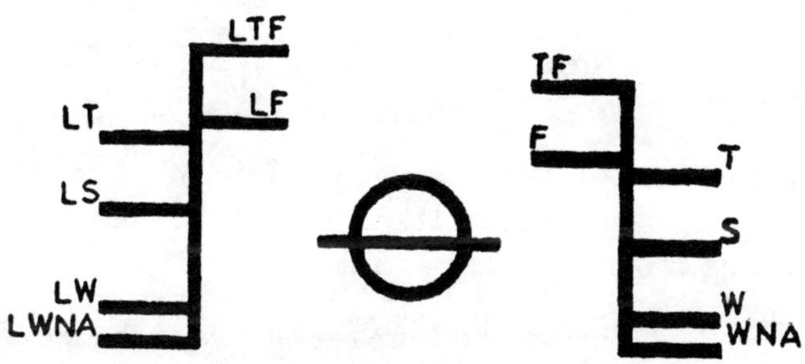

FIG. 4

LT—The "tropical timber load-line"—is placed above the summer timber load-line, at a distance of one forty-eighth of summer draft.

LW—The "winter timber load-line"—is placed below the summer timber line at a distance equal to one thirty-sixth of summer draft.

LWNA—The "winter North Atlantic timber load-line"—is placed level with the ordinary *WNA* line in ships of 100 metres or less in length: or level with the winter load-line in longer ships.

LF—The "fresh water timber load-line" found by allowing the fresh-water allowance above the *LS* line.

LTF—The "tropical fresh-water timber load-line"—found as above, but measured from the *LT* line.

CHAPTER 2

AREAS AND VOLUMES

Areas of Plane Figures.—The areas of certain common plane figures are often used in stability calculations.

Square.—Where a is the length of each side:—

$$\text{Area} = a^2$$

Rectangle.—Where a and b are the lengths of the sides:—

$$\text{Area} = a \times b$$

Triangle.—Where a, b and c are the lengths of the respective sides; h the perpendicular height; and ø the angle between a and b:—

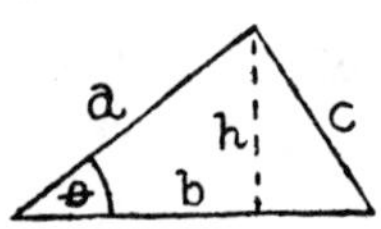

FIG. 5

$$\text{Area} = \frac{bh}{2}$$

$$\text{Area} = \frac{ab \times \sin ø}{2}$$

$$\text{Area} = \sqrt{s\ (s-a)\ (s-b)\ (s-c)}$$

$$\text{where } s = \tfrac{1}{2}\,(a + b + c)$$

Trapezium, or Trapezoid.—There is some disagreement as to the meanings of these two words, but they are often taken to have the same meaning: a four-sided figure, having two of its sides parallel.

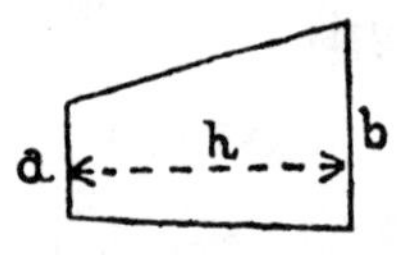

FIG. 6

Where a and b are the lengths of the parallel sides, and h the perpendicular distance between them:—

$$\text{Area} = \frac{h}{2}\ (a + b)$$

Circle.—Where r is the radius; and where π is equal to 3·1416, or approximately $\frac{22}{7}$:—

$$\text{Area} = \pi r^2$$

$$\text{Circumference} = 2\pi r$$

Surface Areas and Volumes.—The following are often useful:—

Cube.—Where a is the length of each edge:—

$$\text{Surface area} = 6a^2$$

$$\text{Volume} = a^3$$

Box shapes.—Where a, b and l are the lengths of the edges:—

$$\text{Surface area} = 2\,(a\,l + b\,l + a\,b)$$
$$\text{Volume} = a\,b\,l$$

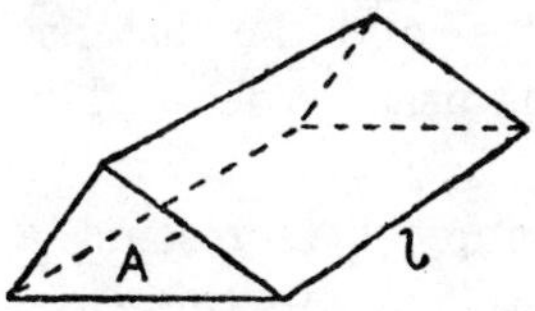

FIG. 7

Wedges and prisms.—Where A is the area of either end and l the length:—

$$\text{Volume} = A\,l$$

Sphere.—Where r is the radius:—

$$\text{Surface area} = 4\pi r^2$$
$$\text{Volume} = \frac{4\pi r^3}{3}$$

Hollow Sphere.—Where r is the internal radius and R the external radius:—

$$\text{Volume of material} = \frac{4\pi R^3}{3} - \frac{4\pi r^3}{3}$$
$$\frac{4\pi}{3}\,(R^3 - r^3)$$

Cylinder.—Where r is the radius and l the length:—

$$\text{Surface area} = 2\pi r\,(r + l)$$
$$\text{Volume} = \pi r^2 l$$

Hollow round section.—Where R is the external radius, r the internal radius and l the length:—

$$\text{Area of cross section} = \pi R^2 - \pi r^2$$
$$= \pi\,(R + r)\,(R - r)$$
$$\text{Volume of material} = \pi l\,(R + r)\,(R - r)$$

Areas of Waterplanes and Other Ship-Sections.—These cannot usually be found with any degree of accuracy by simple mensuration, but there are several methods which may be used to find them. We need only concern ourselves here with "Simpson's Rules" and the "Five-Eight Rule".

Simpson's Rules were designed for finding the area under two types of parabolic curve. These curves are similar to the shapes of the edges of waterplanes and other ship-sections, so we can use them to find areas and volumes of ship-shapes with sufficient accuracy for practical purposes. For this, the First Rule is usually more accurate than the Second Rule and should be used wherever possible.

The preliminary steps in calculating the area of a waterplane or section are as follows. A number of equidistant points are taken along the centre line and perpendiculars are dropped from these points to meet the curved sides. The lengths of these perpendiculars are measured and also the distance between them. The perpendiculars are called "Ordinates" and the distance between them, the "Common Interval". The latter is usually denoted in formulae as "h".

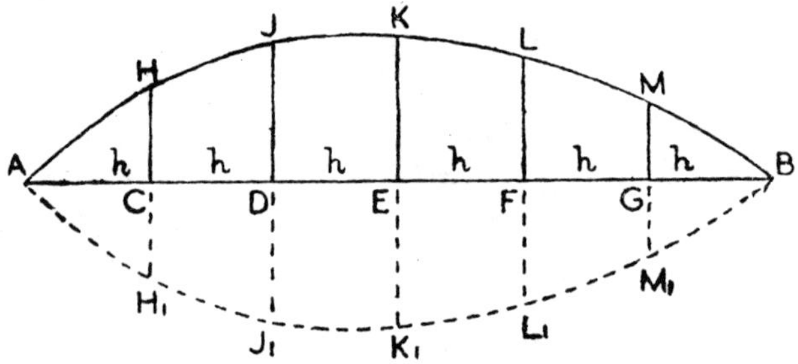

FIG. 8

Figure 8 represents a waterplane. In this case, the centre line (AB) is divided into six parts, each having a length of h (the common interval). The ordinates are HH_1 JJ_1, KK_1, etc. A and B are also ordinates, although in this case they have no length.

It will be noticed that half of the figure has been drawn in plain lines and half in dotted lines. The perpendicular distances shown in the plain lines (CH, DJ, EK, etc.) for the half-waterplane are usually called "Half Ordinates", in order to distinguish them.

When a ship's plans are drawn, they usually shew only the half-waterplane. It is easier, in practice, to measure the half-ordinates from the plans, to put them through the Rules as such, and then to double the half-area so found to give the whole area.

The half ordinates, put through the rules, give the area of the half-waterplane; the ordinates will give the area of the whole waterplane, when put through the same rules.

Simpson's First Rule.—In its simplest form, this rule can be stated:—The area between any three consecutive ordinates is equal to the sum of the end ordinates, plus four times the middle ordinate, all multiplied by one-third of the common interval.

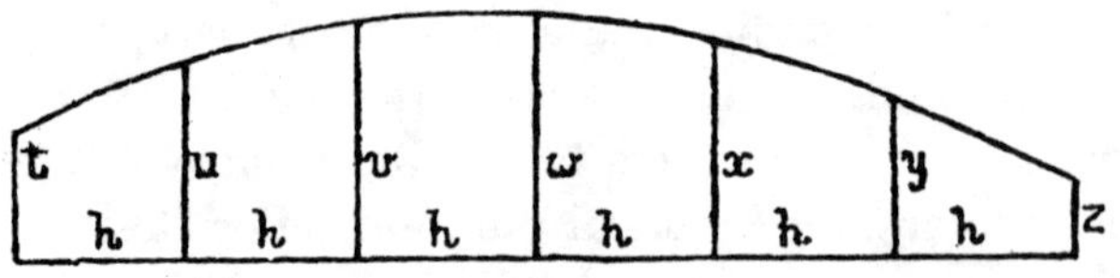

FIG. 9

Consider in Fig. 9 the area contained between the half-ordinates t and v. If the common interval is h, this area equals:—

$$\frac{h}{3}(t + 4u + v)$$

The total area of the half-waterplane can be obtained by finding, in the same way, the areas between v and x, and x and z, and taking the sum of the three.

$$\text{Area between } t \text{ and } v = \frac{h}{3}(t + 4u + v)$$

$$\text{,, ,, } v \text{ and } x = \frac{h}{3}(v + 4w + x)$$

$$\text{,, ,, } x \text{ and } z = \frac{h}{3}(x + 4y + z)$$

$$\text{Total area} = \frac{h}{3}(t + 4u + v) + \frac{h}{3}(v + 4w + x) + \frac{h}{3}(x + 4y + z)$$

$$= \frac{h}{3}(t + 4u + 2v + 4w + 2x + 4y + z)$$

The numbers by which the successive half ordinates, or ordinates, are multiplied (in this case 1, 4, 2, 4, 2, 4, - - - - 1) are called "Simpson's Multipliers".

From the above, we can see two things about this rule:—

(*a*) *It can be used when, and only when, an odd number of ordinates are taken.*

(*b*) *The area is found by multiplying successive ordinates, including the ends, by the multipliers* 1, 4, 2, 4, 2, 4, - - - - 1, *adding the results together, and then multiplying by one-third of the common interval.*

Example.—Use Simpson's First Rule to find the area of a waterplane which has the following half-ordinates, spaced 12 metres apart:—1·7; 5·9; 7·0; 5·2; 1·3 metres.

Since there are 5 ordinates we can use Simpson's First Rule.

$$\text{Half area} = \frac{h}{3}\left\{a + 4b + 2c + 4d + e\right\}$$

$$= \frac{12}{3}\left\{1{\cdot}7 + (4 \times 5{\cdot}9) + (2 \times 7{\cdot}0) + (4 \times 5{\cdot}2) + 1{\cdot}3\right\}$$

Ordinate	Multiplier	Product
1·7	1	1·7
5·9	4	23·6
7·0	2	14·0
5·2	4	20·8
1·3	1	1·3
		61·4

$$\text{Half area} = \frac{12}{3} \times 61{\cdot}4$$

$$\text{Area} = 2 \times \frac{12}{3} \times 61{\cdot}4$$

$$= 491{\cdot}2 \text{ square metres}$$

Simpson's Second Rule.—The area between any four consecutive ordinates is equal to the sum of the end ordinates, plus three times each of the middle ordinates, all multiplied by three-eighths of the common interval.

Consider in Fig. 9, the area between the half-ordinates t and w. If the common interval is h, this area is equal to:—

$$\tfrac{3}{8}h\,(t + 3u + 3v + w)$$

The area between w and z can be found in the same way and added to the above, to give the total area of the half-waterplane.

$$\text{Area between } t \text{ and } w = \tfrac{3}{8}h\,(t + 3u + 3v + w)$$

$$\text{,, \quad ,, \quad } w \text{ and } z = \tfrac{3}{8}h\,(w + 3x + 3y + z)$$

$$\text{Total area} = \tfrac{3}{8}h\,(t + 3u + 3v + 2w + 3x + 3y + z)$$

The conclusions we can draw for this rule are:—

(*a*) *The Rule can be used when, and only when, the number of ordinates is four, or four plus some multiple of three* (e.g., 4, 7, 10, 13, 16, etc.).

(*b*) *The area is found by multiplying successive ordinates, including the ends, by the multipliers* 1, 3, 3, 2, 3, 3, 2, - - - - 1, *adding the results together, and then multiplying by three-eighths of the common interval.*

Example.—Find the area of a waterplane, using Simpson's Second Rule. The common interval is 15 metres and the ordinates are:—0; 9·0; 13·3; 14·7; 12·8; 7·5; 0·6 metres.

$$\text{Area} = \tfrac{3}{8}h\,(a + 3b + 3c + 2d + 3e + 3f + g)$$

Ordinate	Multiplier	Product
0·0	1	0·0
9·0	3	27·0
13·3	3	39·9
14·7	2	29·4
12·8	3	38·4
7·5	3	22·5
0·6	1	0·6
		157·8

$$\text{Area} = \tfrac{3}{8} \times 15 \times 157{\cdot}8$$
$$= 887{\cdot}6 \text{ square metres}$$

The "Five-Eight Rule."—This may be used to find the area between two consecutive ordinates. We must know the length of one other ordinate, next to the area which we wish to measure.

The rule is that *the area is equal to five times one end ordinate, plus eight times the other, minus the known external ordinate (in that order), all multiplied by one-twelfth of the common interval.*

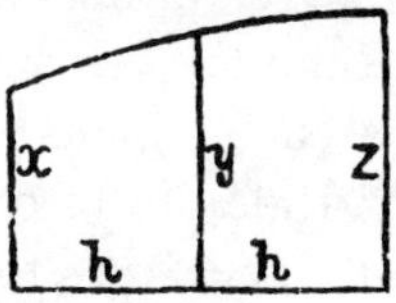

FIG. 10

Suppose that we wish to find the area between x and y in Fig. 10. z is the external ordinate and h the common interval.

$$\text{Then the area} = \frac{h}{12}(5x + 8y - z)$$

Similarly, the area between y and z would be equal to:—

$$\frac{h}{12}(5z + 8y - x)$$

Example.—Three ordinates, a, b and c, are 12 metres apart and have lengths of 29·2, 33·5 and 37·6 metres, respectively. Find the area contained between a and b.

$$\begin{aligned}\text{Area} &= \frac{h}{12}(5a + 8b - c)\\ &= \frac{12}{12}(5 \times 29{\cdot}2 + 8 \times 33{\cdot}5 - 37{\cdot}6)\\ &= 376{\cdot}4 \text{ square metres}\end{aligned}$$

Sharp-ended Waterplanes.—In the above rules, the ends of the waterplanes have been considered as squared-off, but if they are pointed, the rules still apply. The end ordinates are then taken as 0, but are put through the multipliers in the ordinary way.

Unsuitable Numbers of Ordinates.—It sometimes happens that a number of ordinates must be used which will not respond to any of the above rules. In this case, the area is found in two parts, which are later added together.

For example, if there were eighteen ordinates, neither of Simpson's Rules would give the area directly. We could in this case find the area within the first nine ordinates by the First Rule, then add to it the area within the remaining ten (remember that the ninth ordinate would be taken twice) found by the Second Rule. Alternatively, we could find the area between the first seventeen ordinates by the First Rule and that between the remaining two ordinates by the Five-Eight Rule, later adding them together.

For example, if there were eight ordinates, as shewn in Fig. 11, neither the First nor the Second Rule would give the area directly. In this case, we could find, say, the area between *a* and *g* (seven ordinates) by the First Rule; the remaining area, between *g* and *h*, by the Five-Eight Rule. Alternatively, we could use, say, the Second Rule to find the area between *a* and *d* (four ordinates); then the First Rule to find the area between *d* and *g* (five ordinates). The following example will illustrate this.

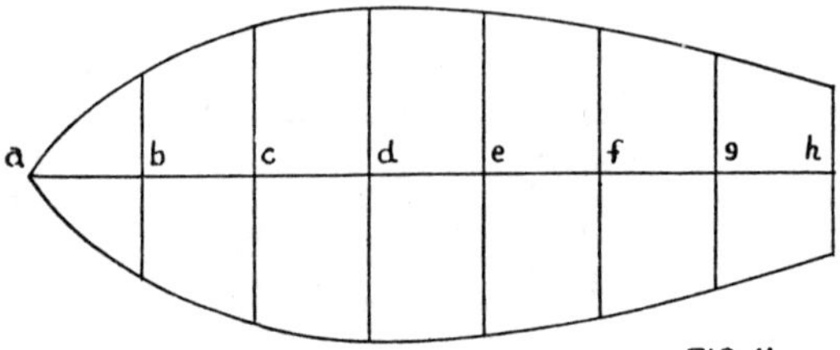

FIG 11

Example.—A waterplane has ordinates, spaced 5 metres apart, of lengths 0; 3·1; 4·9; 6·3; 6·1; 4·8; 2·9; 0·7 metres. Find its area.

First Method:—

Ord.	Multr.	Prod.
0	1	0
3·1	4	12·4
4·9	2	9·8
6·3	4	25·2
6·1	2	12·2
4·8	4	19·2
2·9	1	2·9
		81·7

Ord.	Multr.	Prod.
0·7	5	3·5
2·9	8	23·2
4·8	–1	–4·8
		21·9

First part area $= \frac{5}{3} \times 81{\cdot}7 = 136{\cdot}2\text{m}^2$
Second part area $= \frac{5}{12} \times 21{\cdot}9 = 9{\cdot}1\text{m}^2$

Total area $= 145{\cdot}3\ \text{m}^2$

Second Method:—

Ord.	Multr.	Prod.
0	1	0
3·1	3	9·3
4·9	3	14·7
6·3	1	6·3
		30·3

Ord.	Multr.	Prod.
6·3	1	6·3
6·1	4	24·4
4·8	2	9·6
2·9	4	11·6
0·7	1	0·7
		52·6

First part area $= \frac{3}{8} \times 5 \times 30{\cdot}3 = 56{\cdot}8\text{m}^2$

Second part area $= \frac{5}{3} \times 52{\cdot}6 = 87{\cdot}7\text{m}^2$

Total area $= 144{\cdot}5\text{m}^2$

Note the small discrepancy in the answers above, which is due to differences of accuracy in the different rules used.

Volumes of Ship Shapes.—The ship is divided up into a number of equally spaced sections, the area of each of which is found by Simpson's Rules. The volume is found by putting these areas through the rules in the same way as ordinary ordinates. The sections may be either vertical or horizontal, as convenient. When great accuracy is required, the volume may be found by both methods, one being used to check the other.

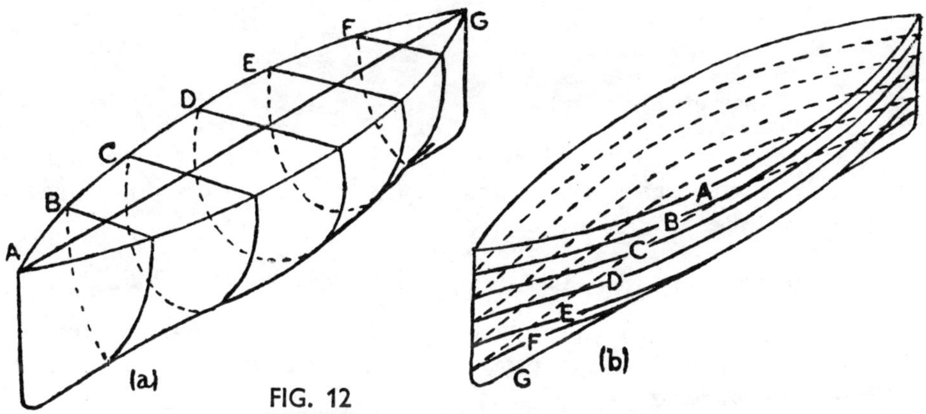

FIG. 12

Fig. 12a shows how the calculation may be done by using the vertical sections B, C, D, E and F. The area of each section is found by the Rules in the ordinary way and the volume can then be calculated by Simpson's First Rule, thus:—

$$\text{Volume} = \frac{h}{3}(A + 4B + 2C + 4D + 2E + 4F + G)$$

Fig. 12b shows how the same volume could be found by using horizontal sections. Where A is the deck area, G the area at the keel, and B, C, D, E and F, the areas of intermediate sections:—

$$\text{Volume} = \frac{h}{3}(A+4B+2C+4D+2E+4F+G)$$

Half-Intervals.—Near the ends of a ship, where the curvature is great, we sometimes bring in intermediate ordinates, halfway between the main ordinates, in order to obtain greater accuracy.

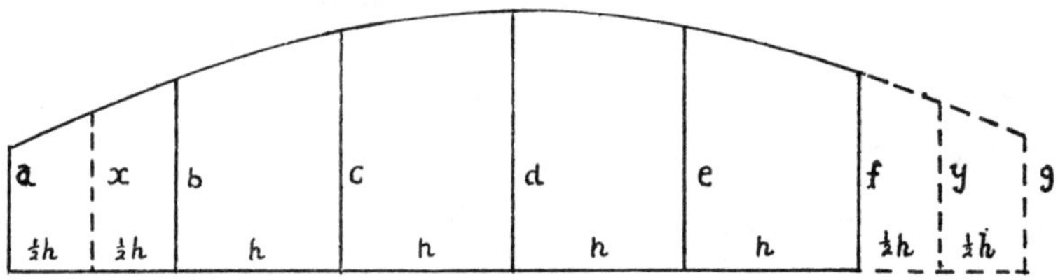

FIG. 13

Consider Fig. 13, which shows a waterplane with an intermediate ordinate midway between a and b. The intervals ax and bx are each $\frac{1}{2}h$.

$$\text{Area between } a \text{ and } b = \frac{\frac{1}{2}h}{3}(a + 4x + b)$$

$$= \frac{h}{3}(\tfrac{1}{2}a + 2x + \tfrac{1}{2}b)$$

$$\text{Area between } b \text{ and } f = \frac{h}{3}(b + 4c + 2d + 4e + f)$$

The areas could be calculated separately and the results added together afterwards. Usually, however, we combine the calculations by adding together the formula, thus:—

$$\text{Total area} = \frac{h}{3}(\tfrac{1}{2}a + 2x + 1\tfrac{1}{2}b + 4c + 2d + 4e + f)$$

If there are intermediate ordinates at the other end of the waterplane, they are treated in the same way. For instance, if we had the ordinate y midway between f and g:—

$$\text{Total area} = \frac{h}{3}(\tfrac{1}{2}a + 2x + 1\tfrac{1}{2}b + 4c + 2d + 4e + 1\tfrac{1}{2}f + 2y + \tfrac{1}{2}g)$$

Example.—A waterplane has ordinates, 20 metres apart, of 1·6, 13·6, 26·5, 25·1, 10·6, and 2·1 metres. There is also an intermediate ordinate, midway between the first two, of 9·4 metres. Find the area.

Ordinate	Multiplier	Product
1·6	½	0·8
9·4	2	18·8
13·6	1½	20·4
26·5	4	106·0
25·1	2	50·2
10·6	4	42·4
2·1	1	2·1
		240·7

$$\text{Area} = \frac{h}{3}(240{\cdot}7)$$

$$= \frac{20}{3} \times 240{\cdot}7$$

$$= 1605 \text{ sq. metres}$$

Coefficient of Fineness of the Midships Section.—This is the ratio between the actual underwater area of a midships section and that of a rectangle of the same depth and width. In Fig. 14 which represents such a section, the coefficient would be:—

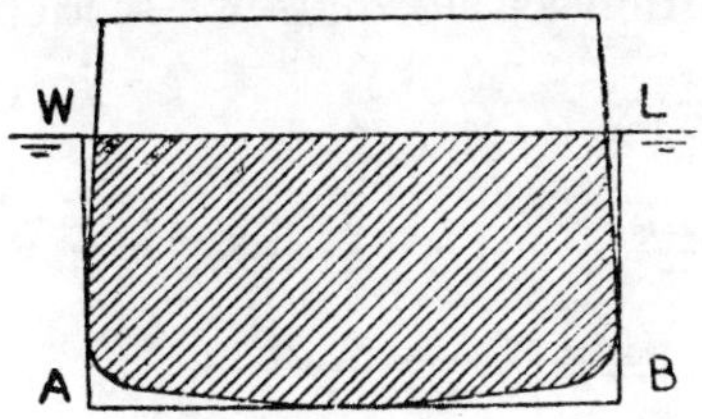

FIG.14

$$\frac{\text{Shaded Area}}{\text{Area of } ABWL}$$

Coefficient of Fineness of a Waterplane.—This is the ratio between the area of the waterplane and that of a rectangle of the same length and breadth. In Fig. 15 the coefficient of fineness of the waterplane shewn is:—

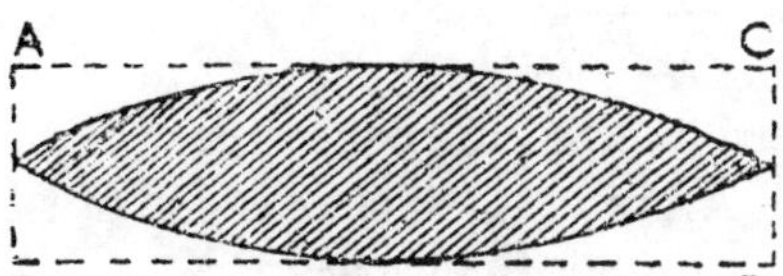

FIG.15

$$\frac{\text{Shaded Area}}{\text{Area of } ABCD}$$

Block Coefficient of Fineness of Displacement.—This is usually referred to by seamen as the "Coefficient of Fineness". It is the ratio between the underwater volume of the ship and that of a box shape, having the same length, breadth and mean draft. In Fig. 16 the shaded volume represents

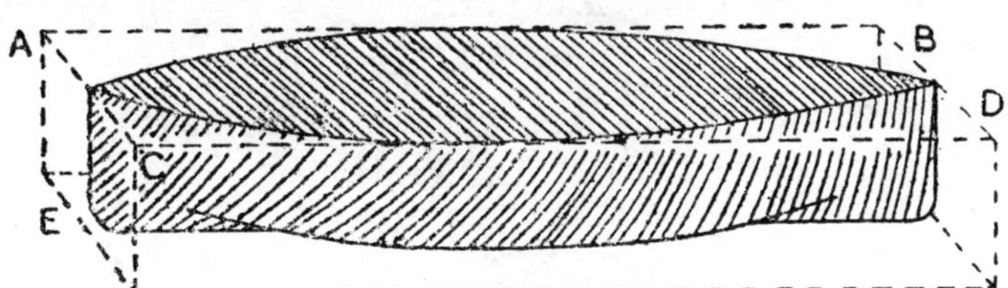

FIG.16 16

the underwater part of a ship and the dotted lines, the box shape mentioned above. The coefficient is then:—

$$\frac{\text{Shaded Volume}}{\text{Volume of } ABCDEFG}$$

Prismatic Coefficient of Fineness of Displacement.—This is the ratio between the underwater volume of the ship and that of a prism having the same length as the ship and the same cross-section as her midships section.

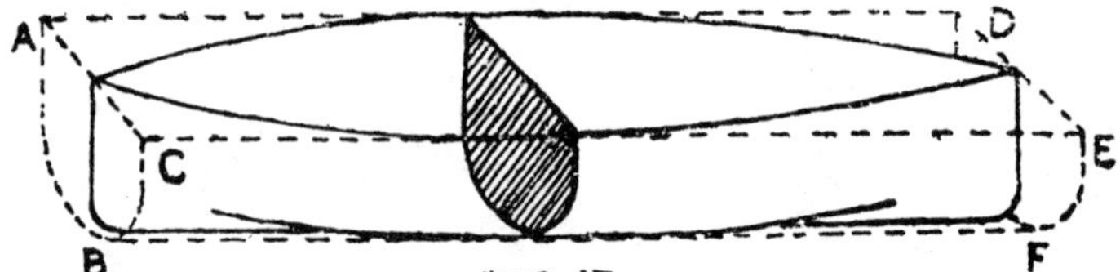

FIG. 17

In Fig. 17, the shaded area represents the underwater part of the midships section; the plain lines, the underwater part of the hull; and the dotted lines, the prism described. The prismatic coefficient of fineness of displacement is then:—

$$\frac{\text{Underwater volume of the ship}}{\text{Volume of the prism } ABCDEF}$$

Wetted Surface.—This is the surface area of the underwater part of the ship's hull. It is of great importance to Naval Architects, since it is one of the factors which resist the movement of the ship through the water

It is almost impossible to calculate the area of wetted surface accurately although it can be found very closely by taking the underwater girths of the ship at regular intervals and then putting the "ordinates" so found through Simpson's Rules. Alternatively, it may be found by approximate formulae, such as:—

$$\text{Wetted surface} = L \{1{\cdot}7d + (C \times B)\}$$

Where L = Length of the ship.
d = Mean draft.
B = Breadth of the ship.
C = Block coefficient of fineness of displacement.

CHAPTER 3

FORCES AND MOMENTS

The stability of ships depends entirely on forces and moments, so that to understand the subject properly, it is necessary to have a general knowledge of such matters.

Force.—This is any push or pull exerted on a body. When a force is being considered, three things must be taken into account:—

1. The amount of force applied.
2. The point at which it is considered to be applied.
3. The direction in which it acts.

A force pushing in one direction has the same effect as an equal force pulling on the opposite side. A point always tries to move directly away from a force pushing at it, or directly towards a force pulling on it. In Fig. 18 the force x, pushing on the point A, has exactly the same effect as an equal force pulling in the direction y. The point will try to move in the direction Ay.

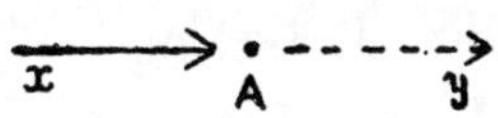

FIG. 18

Resultant Forces.—Any number of forces may act on one point and their combined effect will be the same as that of a single force acting in one particular direction. This imaginary force is called the "resultant force".

If two forces act in one straight line and in the same direction on one point, the resultant force is equal to the sum of the forces and acts in the same direction. In Fig. 19 the forces x and y produce the resultant force R, which is equal to $(x + y)$, which acts in the same direction and which tries to move the point in the direction z.

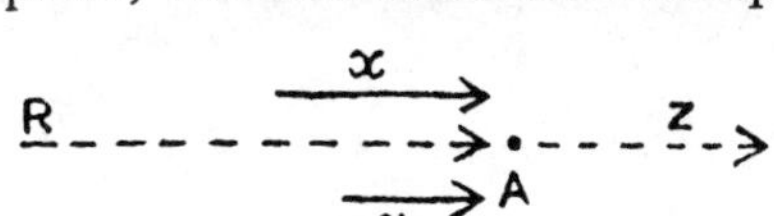

FIG. 19

If two forces act in opposite directions on one point, the resultant force will be equal to the difference of the forces and its direction will be the same as that of the greater of the two. In Fig. 20, x represents the greater of two forces, x and y, acting in opposite directions on the point A. The resultant force will be $(x - y)$, will act in the direction xA and will try to move the point in the direction z.

FIG. 20

When two forces act at an angle to each other, the resultant will depend on the amount and direction of the forces causing it. The resultant is found by using the "Parallelogram of Forces", the principle of which can be seen from Fig. 21. Let x and y be two forces acting on a point A. Let the amounts of the forces be represented by the lengths of AB and AC, respectively. Draw CD parallel to AB and BD parallel to AC. The diagonal AD will represent the amount and direction of the resultant R.

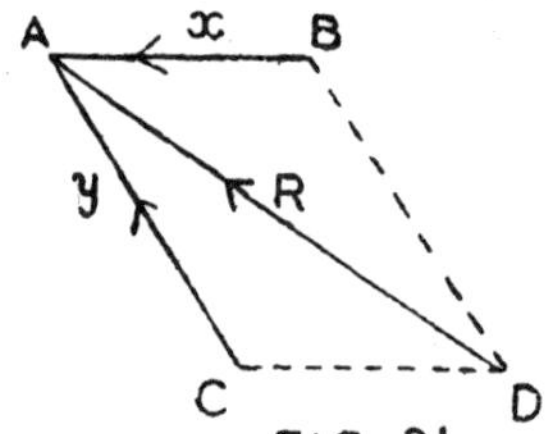

FIG. 21

We sometimes find one force pushing and the other pulling on a point at angles to each other. In such a case, one of the forces must be treated as a push or a pull in the opposite direction, so that both are regarded as pushing, or both pulling. This is allowable, since, as we have seen, a push in one direction is equivalent to an equal pull on the opposite side. Suppose force x to be pulling and force y to be pushing on the point A, as in Figure 22. Force x can be transformed into force z, exerting a push equal to x on the opposite side of A. The parallelogram can then be completed, and the resultant force, R, found in the same way as in the last example.

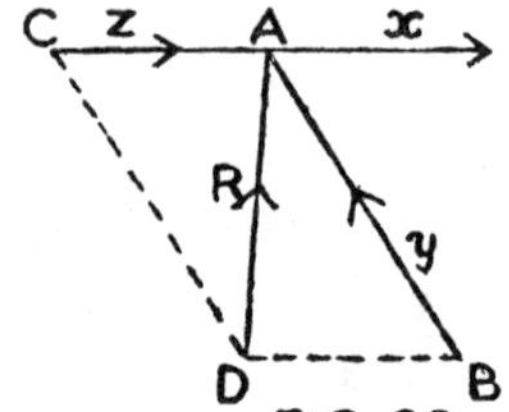

FIG. 22

If measurements are not sufficiently accurate, the resultant can be calculated from the parallelogram by trigonometry.

Moment.—When a body is free to turn about some point and a force is applied to it at some other point, the body will usually try to revolve. Such an attempt on the part of a force to turn a body is called "Moment". Its effect will depend on two things:—

(*a*) The amount of the force.

(*b*) The length of the lever on which the force acts: that is, the perpendicular distance between the direction of the force and the point about which the body can rotate.

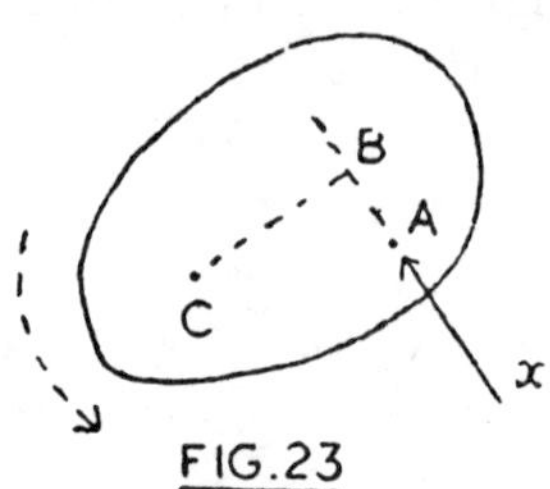

FIG. 23

Suppose that the body shewn in Fig. 23 is free to rotate about the point C, and that a force, x, is applied at the point A. Then the moment trying to turn the body about C is the effect of the force x, acting on the lever BC.

When a lever is used to lift a weight, the "lifting power" depends on the length of the lever and on the amount of force applied to it. If either of these is increased, the lifting power is also increased and vice versa—in other words, the power of a lever is the product of the force and the leverage. Moment is a form of leverage, so we can measure it in a similar way:—

$$\textit{Moment} = \textit{Force} \times \textit{length of lever}.$$

Various quantities are used to measure moments. In ship-stability, where the forces are large, the force is usually measured in tonnes, the lever in metres and the moment in "Tonne-metres".

Properties of a Moment.—Moments can be taken about an imaginary turning point, or about one side of an area, even though the body is not actually free to turn about that point or side. In other words, we could say that if the body were free to turn about a certain point or side, the moment would be so much. Consider a force of x tonnes, applied at a point P, in a direction perpendicular to the surface of the area $KLMN$. If we assume that the area were free to turn about the point C, the moment is $x \times PC$ tonne metres. If we assume that the area can turn about the side KN, the moment about the side is $x \times PB$ tonne-metres, where PB is the perpendicular distance from P to KN. Similarly, the moment about KL would be $x \times PE$ tonne-metres.

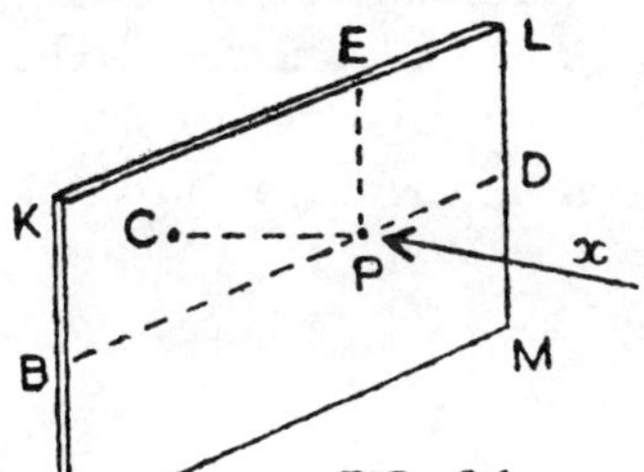

FIG. 24

This property of moments is very important in ship-stability, as we shall see later.

Combinations of Moments.—When several moments are considered to act on one point or on one side of an area, their effects may be combined and considered as that of a single moment. This resultant moment must, of course, be regarded as if it were produced by a different force and/or lever. For this purpose, we take the sum of moments which act in one direction, or the difference of moments which act in opposite directions. This is best seen from examples:—

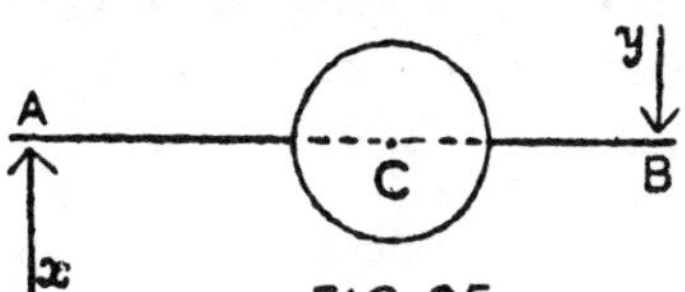

FIG. 25

Consider two men working at a capstan. Let them push with forces of x and y kilogrammes respectively, on the capstan bars AC and BC, as in Fig. 25. Then the moment produced by the first man to turn the capstan about its centre, C, is $x \times AC$ kilo-metres. That produced by the second man is $y \times BC$

kilo-metres. Since they are both trying to turn the capstan in the same direction, the total moment to turn it will be $(x \times AC) + (y \times BC)$ kilo-metres.

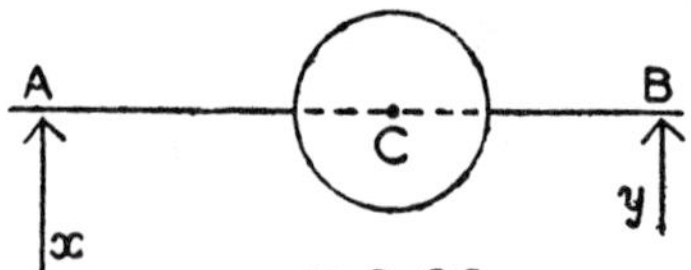

FIG. 26

Now, suppose that the man at B were to turn around and push with the same force, but in the opposite direction; that is, against the man at A. The total moment to turn the capstan has now become $(x \times AC) - (y \times BC)$ kilo-metres. (Fig. 26.)

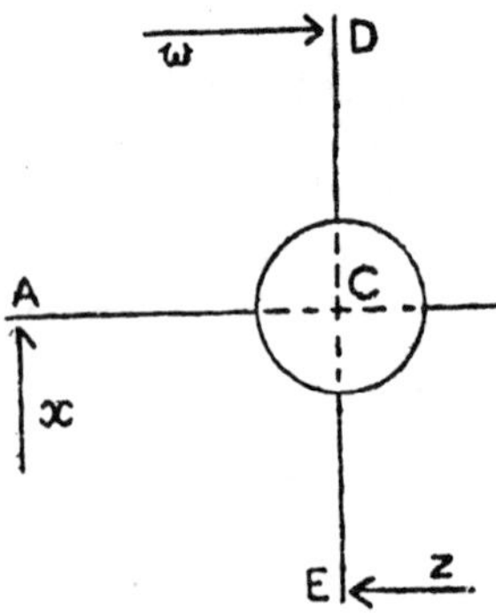

FIG. 27

Suppose that there are four men working at the same capstan, three pushing in one direction and one in the other, against them. (Fig. 27.) The moments in one direction are $(x \times AC)$, $(w \times DC)$ and $(z \times EC)$ kilo-metres. The moment in the other direction is $(y \times BC)$ kilo-metres. The total moment to turn the capstan has become:—

$$(x \times AC) + (w \times DC) + (z \times EC) - (y \times BC)$$
kilo-metres.

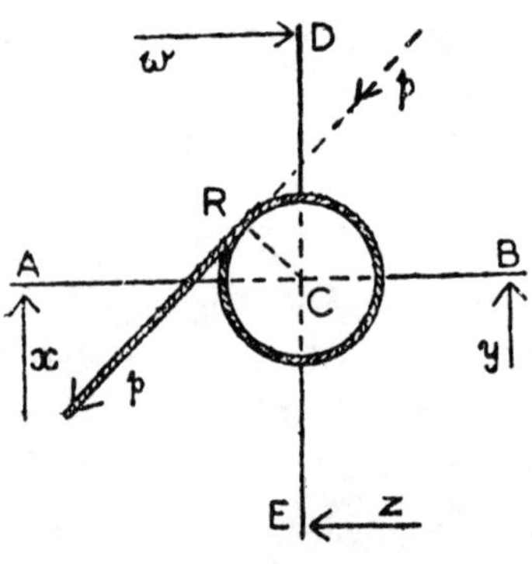

FIG. 28

Finally, let us go one step further and consider the same capstan and men as in the last example, but with a rope around the capstan barrel, pulling against the men with a force of p kilogrammes, as in Fig. 28. The moments trying to turn the capstan in one direction are now $(x \times AC)$, $(w \times DC)$ and $(z \times EC)$ kilo-metres. In the other direction they are $(y \times BC)$ and $(p \times RC)$ kilo-metres. So the total moment to turn the capstan is now:—

$$\{(x \times AC) + (w \times DC) + (z \times EC)\} - \{(y \times BC) + (p \times RC)\} \text{ kilo-metres.}$$

From the above we can see that the *final moment about a point or side of an area is the sum of the moments trying to produce rotation in one direction, minus the sum of the moments trying to produce rotation in the opposite direction.*

Example 1.—A force of 60 kilogrammes is applied to one end of a lever, which is 150 centimetres long. What is the moment about the other end?

Moment = force $\times$ distance = 60×150 = 9000 kilo-centimetres.

Example 2.—Three men are working at a capstan and push on bars 3 metres long with forces of 70, 95 and 86 kilogrammes respectively. Find the moment to turn the capstan.

Moment produced by first man	=	$70 \times 3 = 210$ kilo-metres
,, ,, ,, second man	=	$95 \times 3 = 285$,, ,,
,, ,, ,, third man	=	$86 \times 3 = 258$,, ,,
Total moment to turn capstan	=	753 ,, ,,

Example 3.—A see-saw is exactly balanced about its centre line. Weights of 50 kilogrammes and 90 kilogrammes are placed on one side at distances of 3 metres and 7 metres, respectively, from the centre-line. What weight must be placed on the other side, at a distance of 6 metres from the centre, to cause the see saw to balance once more?

Moment of weights on one side about centre-line $= (50 \times 3) + (90 \times 7)$ tonne-metres.

Let X be the weight on the other side.

Then, moment of weight on the other side, about the centre-line $= 6X$.

For the see-saw to balance, these moments must be equal.

So
$$6X = (50 \times 3) + (90 \times 7)$$
$$X = \frac{(50 \times 3) + (90 \times 7)}{6}$$
$$= 130 \text{ kilogrammes}$$

Couple.—Two equal and opposite forces which act at different points on a body are called a "couple". The moment produced equals the product of one of the forces and the perpendicular distance between them.

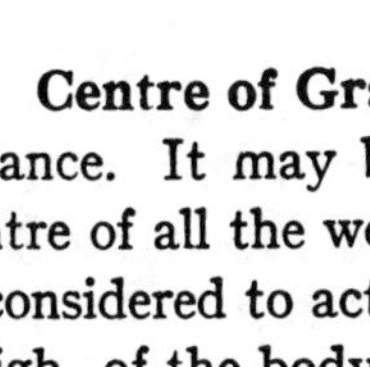

FIG. 29

Moment $= Fd$ tonne-metres

Centre of Gravity.—This is the point about which a body or area would balance. It may be regarded as the geometrical centre of any area, or the centre of all the weight of a body. In the case of a body, the force of gravity is considered to act vertically downwards through it, with a force equal to the weigh of the body.

Centre of Gravity of an Area.—The centres of gravity of certain areas are used in ship-stability, the principal ones being as follows:—

Circle.—The centre of gravity of a circle is ati ts centre.

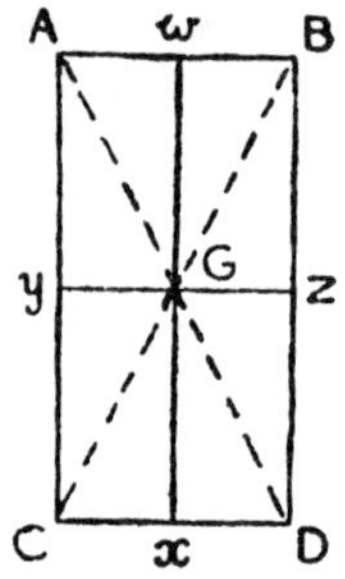

FIG. 30

Square or Rectangle.—The centre of gravity of either of these is at the intersection of the diagonals. When we are considering the stability of box-shaped ships, we always regard the centre of gravity as being at the intersection of the centre-lines. For example, the centre of gravity (*G*) of the rectangle *ABCD*, is at the intersection of the longitudinal centre-line *wx* and the transverse centre-line *yz*.

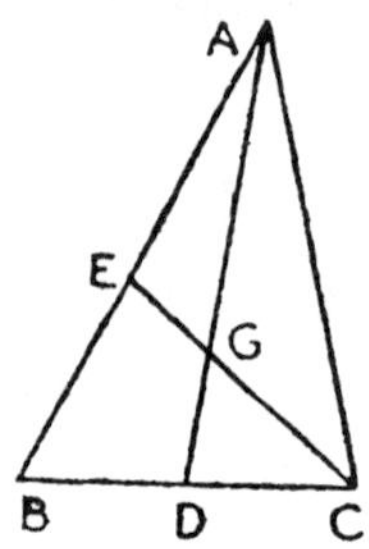

FIG. 31

Triangle. — Draw a median — that is, a line from one corner to bisect the opposite side. The centre of gravity is on this median, at a distance of one-third of its length from the bisected side.

In the triangle *ABC* (Fig. 31), let *AD* be the median, bisecting the side *BC*. Then the centre of gravity will be at *G* and *GD* will equal one-third of *AD*. (Or, $AG = \frac{2}{3}AD$.) Alternatively, draw a second median, *CE*, and *G* will be the point at which *AD* and *CE* intersect.

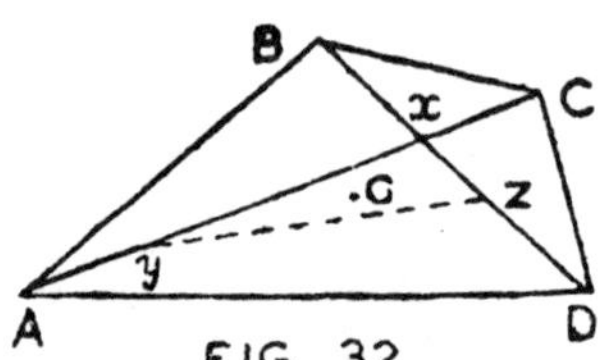

FIG. 32

Trapezium.—Draw the diagonals *AC* and *BD*. Let them intersect at *x*. Measure off *Ay* and *Dz*, so that *Ay* equals *Cx* and *Dz* equals *Bx*. Find the centre of gravity of the triangle *xyz* and this will be the centre of gravity of the trapezium.

Centre of Gravity of a Body.—It is often necessary to find the centre of gravity of a homogeneous body—that is, a solid body, composed of the same material throughout. The three types of body, for which this is usually required are box shapes, prismatic wedges and ship shapes.

For convenience of calculation, we always consider the position of the centre of gravity of such bodies in the transverse and longitudinal directions separately: that is, as it would be seen from the end or from the side respectively. Fig. 33 illustrates this. Let *G* be the centre of gravity of the box shape *ABCDEFH* shown in Fig. (*a*). If we were considering this, we should first find *G* transversely by looking directly at the end, *ABCD*, as in Fig. (*b*) and considering its position in

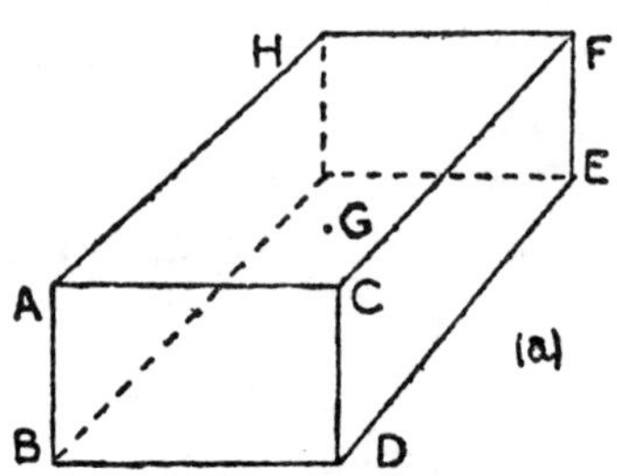

FIG. 33

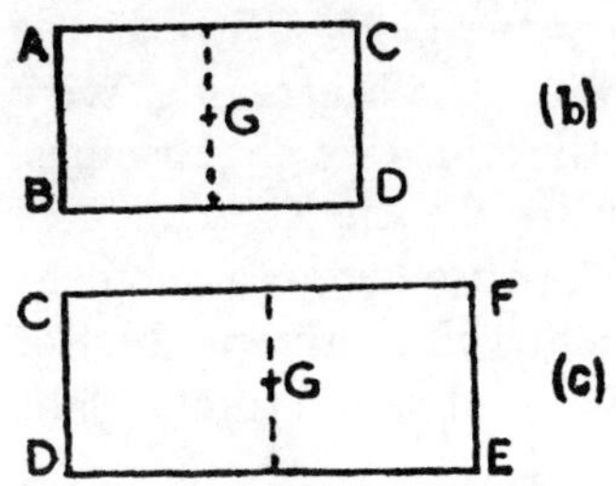

FIG. 33

the plane of that area. We should then find the longitudinal position of *G*, by looking directly at the side *CDEF*, as illustrated in Fig. (*c*).

Box shapes.—We can see from the last figure, that the centre of gravity of a homogeneous box shape is at the centre-line and at the half depth of the body, both longitudinally and transversely.

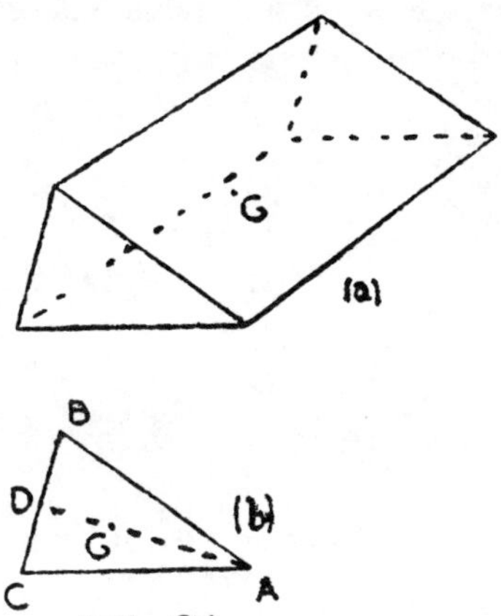

FIG. 34

Prismatic Wedges.—Consider the prismatic wedge shown in Fig. 34 (*a*). In a transverse direction, the position of the centre of gravity will be that of the triangle *ABC*. In other words, it will be on the median *AD* and at a distance of two-thirds of *AD* from the point *A*. Longitudinally, it will be at the half-length, and at the same height as in the transverse direction.

Effect of Added Weights on the Centre of Gravity.—Consider a see-saw, consisting of a plank balanced over a block of wood. When the block is directly under the centre of gravity of the plank (*G*), the see-saw will balance.

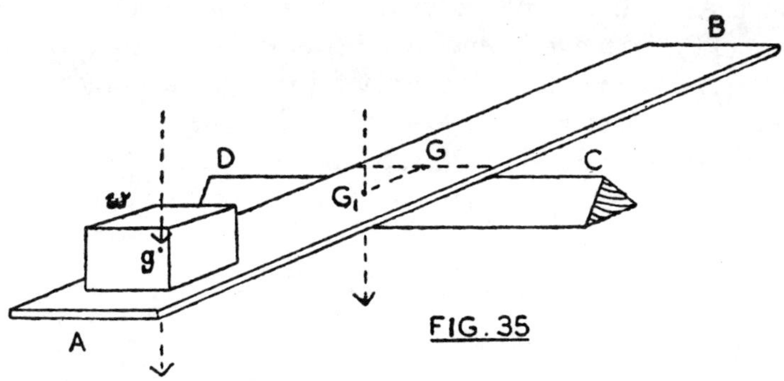

FIG. 35

If a weight, *w*, were now placed on the end *A*, that end would move downwards. We should probably say that this happens because the end *A* has become heavier than the other.

Let us see what has really happened here. We regard the force of gravity as acting vertically downwards through the centre of gravity of a body. Before the weight was added, this force acted directly over the block CD, and the plank balanced because there was no moment to tip it. The centre of gravity of a body is the centre of all the weight in it and since the added weight becomes, in effect, a part of the plank, G moves along to G_1, toward the end A. The force of gravity now acts vertically downward through G_1, producing a moment to cause the end A to move downwards. This moment will be the product of all the weight in the body and of the lever GG_1, so that if W is the total weight after w has been added:—

$$\text{The moment} = W \times GG_1.$$

Since the see-saw balanced before the weight was added, the above effect must be entirely due to the moment of w about G. So we can also say that if g is the centre of gravity of the weight, w:—

$$\text{The moment} = w \times Gg$$
$$\text{and thus } W \times GG_1 = w \times Gg.$$

From this, we can derive a formula for the shift of the centre of gravity of a body when a weight is added to it:—

$$GG_1 = \frac{w \times Gg}{W}$$

Two things are obvious from this:—

(*a*) *The centre of gravity of a body will move directly towards the centre of gravity of any weight added to it.*

(*b*) *The distance it will move will equal the moment of the added weight about the centre of gravity of the body, divided by the total weight after the weight has been added.*

The Effect of Removing Weights.—Suppose that in the last example, we had sawn a piece off the end, A, of the see-saw, instead of adding weight. The end B would then move downwards, because the centre of gravity of the plank would move towards it. If we use a similar process of reasoning to that which we used in the last section, we shall see that the moment of the weight removed must, again, be equal to the total moment caused by the shift of the centre of gravity of the plank. Thus, if W is the total weight after the piece has been sawn off:—

$$GG_1 = \frac{w \times Gg}{W}$$

Two things are again obvious:—

(*a*) *The centre of gravity of a body will move directly away from the centre of gravity of any weight removed from it.*

(*b*) *The distance it will move will equal the moment of the weight removed about the centre of gravity of the body, divided by the total weight remaining after the weight has been removed.*

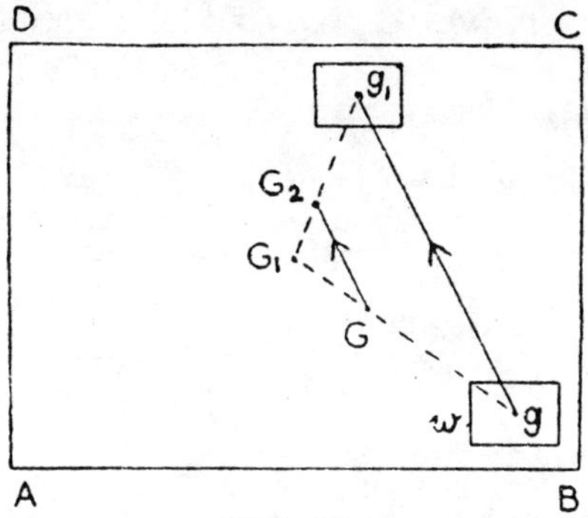

FIG. 36

The Effect of Shifting Weights.—Consider a flat plate *ABCD*, having a weight *w* placed on it. Let *g* be the centre of gravity of this weight, and *G* the centre of gravity of plate and weight together. Let *W* be the total weight of the whole mass.

If we remove the weight from the plate, *G* will move directly away to G_1. If we then replace the weight so that its centre of gravity is at g_1, the centre of gravity of the whole will now move from G_1 to G_2, in the direction G_1g_1.

We can shew, by working out the moments, that:—

(*a*) gg_1 is parallel to GG_2

(*b*) $W \times GG_2 = w \times gg_1$

We can say from the above that:—

(*a*) *The centre of gravity of a body moves parallel to the shift of the centre of gravity of any weight shifted within it.*

(*b*) *The distance it moves is equal to the weight shifted, multiplied by the shift of its centre of gravity, all divided by the total weight of the body.*

The Use of Moments to find the Centre of Gravity of a Body.—It has already been shewn that for a plank of weight *W* kilos and a line *AB*:—

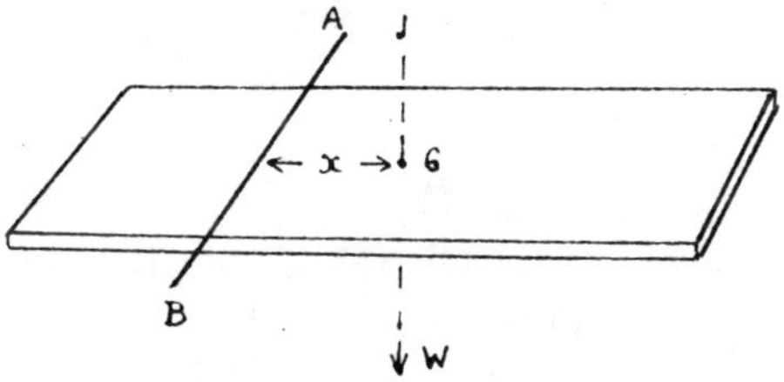

FIG. 37

Moment about $AB = Wx$ kilo-metres

Conversely, if we know, or can find the weight and moment of the plank about *AB*, we can find *x*, since

$$x = \frac{\text{Moment about } AB}{W}$$

The Use of Moments to find the Centre of Gravity of an Area.—Areas are considered to have moments, but this is more difficult to visualise because an area has no actual weight. However, if we think of an area as a very thin "plank", of infinitesmal weight, we could then imagine it as having a very small moment and the above formula would apply to it. In this case, there will be unit weight per unit area: that is, each square metre of the area will weigh, say, *p* grammes.

Total weight of area $= p \times$ area in square metres.

Now, if the plank in the last section were so thin as to become an area only the position of its centre of gravity, relative to AB could be found, as follows:—

Weight of plank (W) = Area of plank $\times p$

Moment of plank = Moment of area about $AB \times p$

Now, $$x = \frac{\text{Moment of plank about } AB}{W}$$

So, $$x = \frac{\text{Moment of area about } AB \times p}{\text{Area of plank} \times p}$$

p cancels out, leaving:—

$$x = \frac{\text{Moment of area about } AB}{\text{Area of plank}}$$

To Find the Centre of Gravity of a Waterplane.—The position of the centre of gravity of a waterplane can be found from the above formula if its area and its moment about one end or one ordinate are calculated.

The area can be found by means of Simpson's Rules, as shewn in Chapter 2.

The moment could be found by dividing the waterplane into a series of narrow transverse strips. The area of each strip multiplied by its distance from the chosen ordinate will give its moment: and the moment of the whole waterplane will be the algebraic sum of the moments of all the strips.

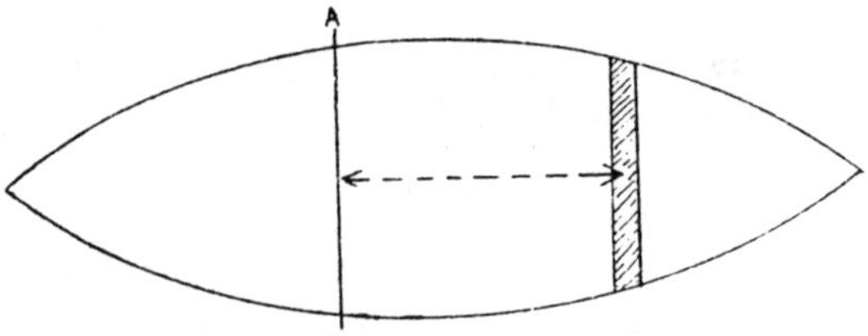

FIG. 38

For example, in Fig. 38, the moment of the strip shewn about AB would be:—

Area of strip $\times$ mean distance from AB

If the whole waterplane were divided into a series of such strips:—

Moment of waterplane = Algebraic sum of moments of all strips, i.e. Sum of all moments to the right of $AB \sim$ sum of all moments to the left of AB.

Distance of centre of gravity of waterplane from AB: =

$$\frac{\text{Moment of waterplane}}{\text{Area of waterplane}}$$

The centre of gravity will lie on that side of AB which has the greater moment.

In practice, the above would be a laborious procedure if it were carried out in full, but fortunately Simpson's Rules provide us with a convenient short method. When these Rules are used to find an area, they divide the waterplane up, in effect, into a number of narrow strips and then give the sum of the areas of all the strips. If, when using the Rules, we multiply the product for each ordinate by its distance from the chosen ordinate, it will give us a product for the moment. The algebraic sum of the products of moments for all the ordinates multiplied by one-third of the common interval will equal the sum of the moments of all the strips: that is, it will give us the moment of the area about the chosen ordinate.

This can best be seen from an example. Let us find the position of the centre of gravity of a waterplane, 250 metres long, which has the following ordinates, spaced 25 metres apart, from forward to aft: 0·4: 7·8: 17·2: 21·1: 27·5: 30·0: 29·3: 28·2: 22·5: 15·6: 1·0 metres. Put these ordinates through Simpson's Rules, as follows, to find the position of the centre of gravity relative to the mid-ordinate.

No. of Ord.	Length	Mult'r	Product for Area	Interval from Mid-Ord.	Product for Moment	Product for Moment
1	0·4	1	0·4	5	2·0	
2	7·8	4	31·2	4	124·8	
3	17·2	2	34·4	3	103·2	
4	21·1	4	84·4	2	168·8	
5	27·5	2	55·0	1	55·0	453·8 for'd
6	30·0	4	120·0	0	0·0	
7	29·3	2	58·6	1	58·6	
8	28·2	4	112·8	2	225·6	
9	22·5	2	45·0	3	135·0	
10	15·6	4	62·4	4	249·6	
11	1·0	1	1·0	5	5·0	673·8 aft
			605·2			220·0 aft

(Strictly speaking, this should be the *distance* from the mid-ordinate but it makes for more simple arithmetic if we take the intervals from the mid-ordinate and then multiply afterwards by the common interval, as shewn.)

$$\text{Area} = \frac{h}{3} \times \text{Product for area} = \frac{25}{3} \times 605{\cdot}2 = 5043{\cdot}3 \text{ m}^2$$

$$\text{Moment about mid-ordinate} = \text{Product for moment} \times \frac{h}{3} \times h$$

$$= 220 \times \frac{25}{3} \times 25 = 45833$$

Distance of centre of gravity from mid-ordinate =

$$\frac{\text{Moment}}{\text{Area}} = \frac{45833}{5043{\cdot}3} = 9{\cdot}09 \text{ metres}$$

The greater moment is abaft the mid-ordinate, so the centre of gravity will be 9·09 metres abaft that ordinate

To Find the Centre of Buoyancy of a Ship Shape.—

In Chapter 2 it has been shewn how we can obtain the volume of a ship shape, by putting cross-sectional areas through Simpson's Rules as if they were ordinates. Similarly, if we put cross-sectional areas through the process described in the last section, we can obtain the position of the centre of gravity of a homogeneous ship shape. The centre of gravity of a ship's underwater volume is the centre of buoyancy. So if we take a series of equally-spaced sections for the ship's underwater volume and put them through the Rules, we shall obtain the fore and aft position of the centre of buoyancy. Similarly, a series of equally-spaced waterplanes, put through the Rules will give the vertical position of the centre of buoyancy.

Example.—A ship's underwater volume is divided into the following vertical cross-sections, from forward to aft, spaced 20 metres apart: 10; 91; 164; 228; 265; 292; 273; 240; 185; 111; 67 square metres. If the same underwater volume is divided into waterplanes, 2 metres apart, their areas, from the keel upwards are: 300; 2704; 3110; 3388; 3597; 3759; 3872 square metres. Find the position of the centre of buoyancy (*a*) fore and aft, relative to the mid-ordinate, (*b*) vertically, above the keel.

No. of Ord.	Area	Mult'r	Product for Volume	Interval	Product for Moment	
1	10	1	10	5	50	
2	91	4	364	4	1456	
3	164	2	328	3	984	
4	228	4	912	2	1824	
5	265	2	530	1	530	4844 for'd
6	292	4	1168	0	0	
7	273	2	546	1	546	
8	240	4	960	2	1920	
9	185	2	370	3	1110	
10	111	4	444	4	1776	
11	67	1	67	5	335	5687 aft
			5699			843 aft

$$\text{Volume} = \frac{h}{3} \times \text{product for volume} = \frac{20 \times 5699}{3} = 37993 \text{ m}^3$$

$$\text{Moment} = \text{product for moment} \times \frac{h}{3} \times h = 843 \times \frac{20}{3} \times 20 = 112400$$

$$\text{Position of centre of buoyancy} = \frac{\text{Moment}}{\text{Volume}} = \frac{112400}{37993} = \begin{array}{l}2{\cdot}96 \text{ m abaft}\\ \text{mid-ordinate.}\end{array}$$

No. of Ord.	Area	Mult'r	Product for Volume	Interval	Product for Moment	
	300	1	300	0	0	
1	2704	4	10816	1	10816	
2	3110	2	6220	2	12440	
3	3388	4	13552	3	40656	
4	3597	2	7194	4	28776	
5	3759	4	15036	5	75180	
6	3872	1	3872	6	23232	
			56990		191100	

$$\text{Volume} = \frac{h}{3} \times \text{product for volume} = 2 \times \frac{56990}{3} = 37993 \text{ m}^3$$

$$\text{Moment} = \text{product for moment} \times \frac{h}{3} \times h = 191100 \times \frac{2}{3} \times 2 = 254800$$

$$\text{Position of centre of buoyancy} = \frac{\text{Moment}}{\text{Volume}} = \frac{254800}{37993} = 6{\cdot}71 \text{ m above keel}$$

The Use of Intermediate Ordinates.—If we are given intermediate ordinates when finding the centre of gravity of an area or volume, we use the same basic method as above. In this case, however, the multipliers are modified as shewn in Chapter 2; whilst the intervals for the moments are measured in intervals and half-intervals. This is best seen from examples.

Example 1.—A waterplane is 60 metres long and has ordinates, from forward to aft of 1·0, 6·9, 11·1, 11·2, 8·9 and 5·0 metres. There is also an intermediate ordinate of 3·8 metres, midway between the first two forward ones. Find the area and position of the centre of gravity of this waterplane.

Let us take moments about the 11·2 metres ordinate.

No. of Ord.	Length	Mult'r	Product for Area	Interval	Product for Moment	
1	1·0	½	0·5	3	1·5	
1½	3·8	2	7·6	2½	19·0	
2	6·9	1½	10·4	2	20·8	
3	11·1	4	44·4	1	44·4	85·7 for'd
4	11·2	2	22·4	0	0	
5	8·9	4	35·6	1	35·6	
6	5·0	1	5·0	2	10·0	45·6 aft
			125·9			40·1 for'd

$$\text{Common interval } (h) = \frac{60}{5} = 12 \text{ metres.}$$

$$\text{Area} = \frac{h}{3} \times \text{product for area} = \frac{12}{3} \times 125{\cdot}9 = 503{\cdot}6 \text{ m}^2.$$

$$\text{Moment} = \text{Product for moment} \times \frac{h}{3} \times h = 40{\cdot}1 \times \frac{12}{3} \times 12 = 1924{\cdot}8$$

$$\text{Position of centre of gravity} = \frac{\text{Moment}}{\text{Area}} = \frac{1924{\cdot}8}{503{\cdot}6}$$

$$= 3{\cdot}8 \text{ metres (forward of } 11{\cdot}2 \text{ m ordinate)}$$

Example 2.—A vessel has waterplanes, 2 metres apart, from the keel upward, of 24, 400, 807 and 1032 square metres. An intermediate waterplane 1 metre above the keel, has an area of 153 square metres. Find the displacement and *KB*, when the vessel is floating at the uppermost waterplane.

Taking the moments about the keel.

No. of Ord.	Area	Mult'r	Product for Volume	Interval	Product for Moment
1	24	½	12	0	0
1½	153	2	306	½	153
2	400	1½	600	1	600
3	807	4	3228	2	6456
4	1032	1	1032	3	3096
			5178		10305

$$\text{Volume} = \frac{h}{3} \times \text{product for volume} = \frac{2}{3} \times 5178 = 3452 \text{ m}^3$$

$$\text{Displacement} = \text{Volume} \times 1{\cdot}025 = 3452 \times 1{\cdot}025 = 3539 \text{ tonnes}$$

$$\text{Moment} = \text{product for moment} \times \frac{h}{3} \times h = 10305 \times \frac{2}{3} \times 2 = 13740$$

$$KB = \frac{\text{Moment}}{\text{Volume}} = \frac{13740}{3452} = 3{\cdot}99 \text{ metres.}$$

Appendages.—An appendage is a small area (or volume) adjacent to the main area (or volume), which has for some reason not been measured into the main one. Its effect on the centre of gravity can be found by applying its area (or volume) and moment about the chosen ordinate, to those for the main area.

For example, suppose that the waterplane given in the example on waterplanes, had an appendage of area 245 square metres, with its centre of gravity 135 metres abaft the mid-ordinate:—

Main area was	..	5043 m^2
Appendage ,,	..	245 m^2
Total area	..	5288 m^2

Moment of main area was	45833 aft
Moment of appendage = area × distance mid-ordinate	
= 245 × 135	33075 aft
Total Moment	78908 aft

Position of centre of gravity of area and appendage, abaft the mid-ordinate =

$$\frac{\text{Moment}}{\text{Area}} = \frac{78908}{5288} = 14{\cdot}9 \text{ metres.}$$

The effect of an appendage on the centre of gravity of a homogeneous ship shape can be calculated in the same way.

Inertia.—A stationary body resists any attempt to move it and a moving body any attempt to change its speed or direction. This property is called "inertia" and a certain amount of force must be exerted to overcome it. If we consider what would happen if we tried to play football with a cannon ball, it should be obvious that the greater the weight of the body, the greater will be its inertia. Thus, the weight of a body gives a measure of its inertia so far as ordinary non-rotational motion is concerned. For the sake of correctness, we shall, from now on, use the word "mass" instead of "weight", but for our present purpose we may take it to mean the same thing.

Moment of Inertia and Radius of Gyration.—It has been shewn earlier in this chapter that in ordinary motion, the behaviour of a body depends on the amounts of the forces applied to it: but that where a turning, or rotational movement is attempted, the behaviour of the body depends on the *moments* of the forces applied. In a somewhat similar way, although the inertia of ordinary motion is governed by mass, the inertia of rotational motion is governed by a quantity called its "moment of inertia", or "second moment". There is this difference, however, that both inertia and moment of inertia are independent of the forces applied to the body. Roughly speaking we may say that in the case of ordinary motion, the greater the mass, or inertia, the greater the resistance of the body to being moved; in the case of rotational

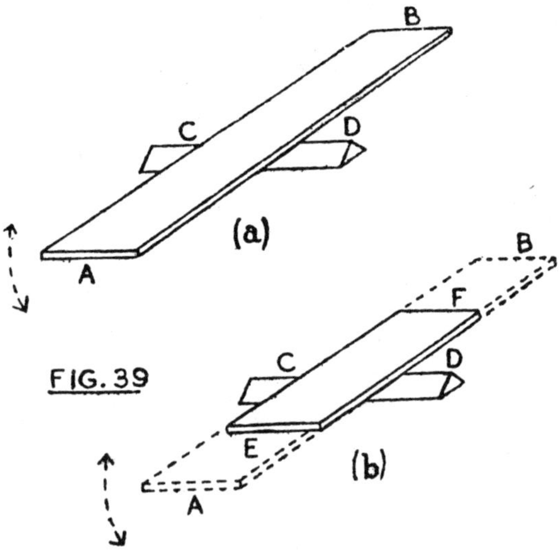

FIG. 39

motion, the greater the moment of inertia, the greater the resistance to rotation.

What is moment of inertia and upon what properties of bodies does it depend? Consider the see-saw shown in Fig. 39 (*a*), consisting of a wooden plank, *AB*, balanced across a block, *CD*. If it were struck a downward blow at *A*, it would start to oscillate up and down. Suppose that the wooden plank were now replaced by an iron one of exactly the same dimensions. Strike this iron "plank" a blow of exactly the same force as before and it will also oscillate, but much less than did the lighter wooden one. This shows that moment of inertia depends on mass—as a matter of fact, it is directly proportional to mass.

Suppose that we now replace the plank by a shorter one of equal weight, as shewn in Fig. 39 (*b*). If this plank is struck a downward blow with the same force as before, it will oscillate quite violently. Since the weight is the same, the increase in the rate of oscillation must be due to the decrease in length, so we can conclude that moment of inertia will also depend on length. Unfortunately, it is not proportional to actual length, but to the square of what we may term the "effective length" of the plank. This effective length is called "radius of gyration" and it can be shown that where l is the actual length:—

$$\text{Radius of gyration} = \sqrt{\frac{l^2}{12}}$$

The discoveries of the last two paragraphs indicate that

$$\text{Moment of inertia} = \text{mass} \times (\text{radius of gyration})^2.$$

Thus, where I is the moment of inertia, m the mass (or weight) of the plank, and l the actual length:—

$$I = \frac{ml^2}{12}$$

If the plank is of unit weight per unit length (i.e., if it had, say, a weight of one kilogramme per metre of length), m and l will be numerically the same (i.e., a 5 metre plank would weigh 5 kilogrammes, so that m and l would both be 5).

For our purpose, we can assume this and substitute l for m in the above formula, which becomes:—

$$I = \frac{l^3}{12}$$

One factor which we have not so far considered is that moment of inertia also depends on the position of the axis about which the body is assumed to rotate. Throughout the above, we have only considered the plank of a see-saw rotating about its centre-line and the above formula only holds good for such rotation. In the case of rotation about other axes, the formula will be modified.

Moment of Inertia of a Body About its Centre-Line.—Our discoveries about the moment of inertia of planks also extend to bodies of any size. The integral calculus is involved in most cases, but roughly speaking, the method of calculating the moment of inertia about the axis of rotation is to divide the body into strips, which may be likened to planks. The total moment of inertia of all the planks will give that of the body. For instance, the moment of inertia of the body shewn in Fig. 40 about its centre-line, AB could be found by dividing it into a number of thin "planks" as shewn. If the moment of inertia of each "plank" were then found, the sum of all of these would be the moment of inertia of the whole body.

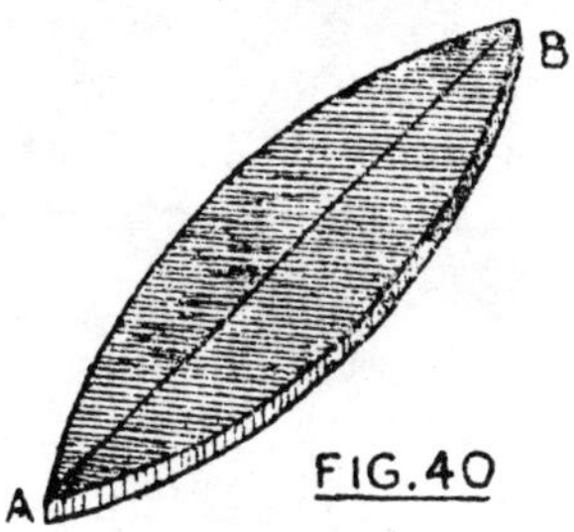

FIG. 40

Moment of Inertia of a Waterplane about its Centre-Line.—If the body shewn in Fig. 40 were infinitely thin, it would resemble a ship's waterplane and it is obvious that the moment of inertia of such a waterplane could roughly be found as described above. For ship shapes, this involves the use of the calculus, but for box shapes, having a rectangular waterplane, the calculation is more simple. Consider such a waterplane, $PQRS$ having the centre-line CD. Let l be its length and b its breadth. If we consider this waterplane to be composed of a number of infinitely thin planks parallel to QR, the moment of inertia will be equal to the sum of the moments of inertia of all the planks. The moment of inertia of each plank about CD will be:—

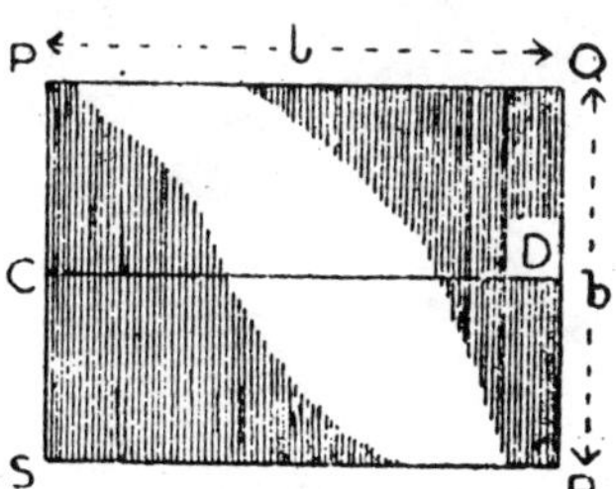

FIG. 41

$$\frac{b^3}{12}$$

Since the planks, placed side by side, extend over the length l we could say that if there were one plank for each unit of length, there would be l planks. The total moment of inertia of l similar planks would be l times that of one plank, so where I is the moment of inertia of the area, l its length and b its breadth:—

$$\text{So, for a rectangular waterplane:—}\quad I = \frac{lb^3}{12}$$

Equilibrium.—This may roughly be defined as the state of balance of a body. There are three states of equilibrium—stable, neutral and unstable—and the body may be in either of these, according to the relative positions of the centre of gravity and the point of support.

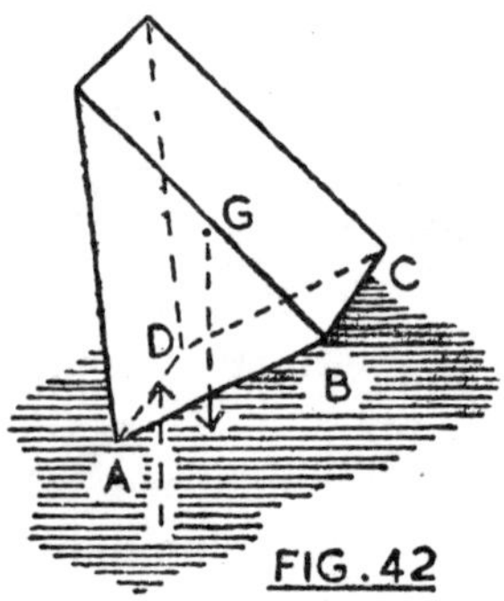

FIG. 42

Consider a wedge which is tilted over on one edge. It will be in stable equilibrium if it tends to return to its original upright position: that is, if the centre of gravity falls vertically over some point in the base, inside the edge on which it is supported.

Fig. 42 shews a wedge tipped over on its edge *AD*, with *G* its centre of gravity and *ABCD* its base. As long as *G* is vertically over some point inside the area *ABCD*, the force of gravity acting vertically downwards through *G* will try to pull the wedge back to its original position. Such a wedge is in stable equilibrium.

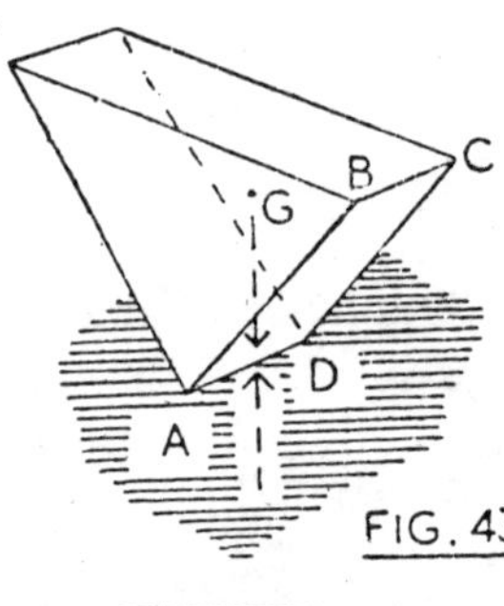

FIG. 43.

The wedge is in neutral equilibrium if it is so balanced that it neither tends to fall over, nor to return to its original position. In this case, the centre of gravity must be vertically over the point of support, as shown in Fig. 43. When *G* is vertically over *AD*, the force of gravity does not tend to turn the wedge either way. It will thus, in theory at least, remain balanced until some other force is applied to it.

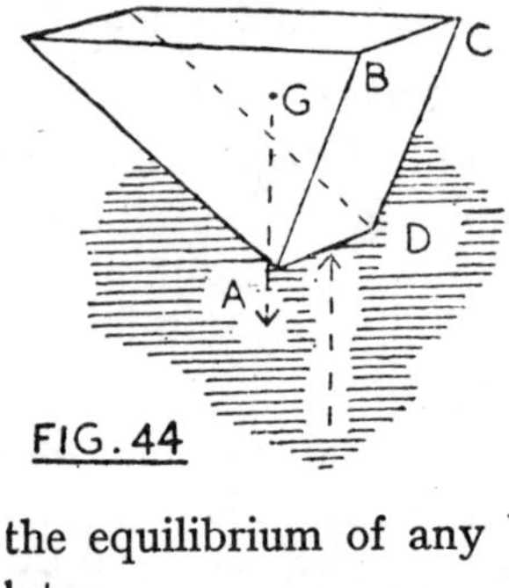

FIG. 44

The wedge is in unstable equilibrium if it tends to fall over on to its side. This will happen if the centre of gravity is vertically over some point outside the base, as is shewn in Fig. 44. If *G* falls vertically outside *AD*, the force of gravity pulling down through *G* will cause the wedge to fall over on to its side.

All the above are general cases, and apply to the equilibrium of any body. The special cases of ships will be considered later.

CHAPTER 4

DENSITY, DEADWEIGHT AND DRAFT

We have already seen that the volume displaced by a floating body varies inversely as the density of the water in which it floats. This means that if a ship's displacement remains the same, her draft will increase if she enters water of less density; or will decrease if she enters water of greater density.

We also know that a ship which loads cargo (i.e. increases her deadweight) will increase her draft, whilst a ship which discharges cargo will decrease her draft, if the density of the water does not change.

The Effect of Density on Draft of Box Shapes.—In the case of *box shapes*, the volume displaced is equal to the product of length, breadth and draft; so we can say:—

$$\frac{\text{New volume displaced}}{\text{Old volume displaced}} = \frac{\text{Old density}}{\text{New density}}$$

$$\frac{\text{Length} \times \text{Breadth} \times \text{New Draft}}{\text{Length} \times \text{Breadth} \times \text{Old Draft}} = \frac{\text{Old density}}{\text{New density}}$$

$$\frac{\text{New Draft}}{\text{Old Draft}} = \frac{\text{Old density}}{\text{New density}}$$

Effect of Density on Draft of Ship Shapes.—These also increase their draft when the density of the water decreases and vice versa, but in this case the change is not in proportion and its calculation is more complicated. We overcome this difficulty by giving each ship a "Fresh Water Allowance" when her load-lines are assigned. This allowance is approximately the amount by which the ship will decrease her draft on going from fresh water to salt water.

The ordinary load-lines show the draft at which a ship can safely remain at sea. In the smooth water of a harbour or river, it would be quite safe to load her a little below these marks, provided that she rises to them when or before she reaches the open sea. A ship loading in a harbour of fresh water could submerge her load lines by the amount of her fresh-water allowance, since she would rise to her proper load line on reaching salt water.

Ships often load in dock water which is brackish—that is, which has a density of more than 1000 and less than 1025 kg/m³. In this case, the amount by which the salt water load-lines can be submerged is found from the fresh-water allowance by simple proportion. This amount is often called the "Dock Water Allowance".

Let x equal the amount by which the load-line can be submerged in water of density, δ. Let Fa equal the fresh-water allowance.

A change of density from 1000 to 1025 kg/m³ will produce a change of draft of Fa centimetres.

A change of density from δ to 1025 will produce a change of draft of x centimetres.

We can assume, for practical purposes, that the change of draft is proportional to the change of density. Hence:—

$$\frac{x}{Fa} = \frac{1025 - \delta}{1025 - 1000}$$

$$x = \frac{Fa\ (1025 - \delta)}{25}$$

Note that where, as in this case, we are dealing with *change* of draft and density, these are directly proportional. In box shapes, when we are dealing with *actual* draft and density, they are inversely proportional.

Example 1.—A ship floats at a draft of 5·42 m in water of density 1·007 t/m³. If her fresh-water allowance is 242 mm, what would be her draft on passing into salt water?

$$\text{Dock water allowance } (x) = \frac{\text{Fa}\ (1025 - \delta)}{25}$$

$$= \frac{242\ (1025 - 1007)}{25}$$

$$= 174\text{mm, or } 0{\cdot}17\text{m (Rise)}$$

$$\text{Old draft} = 5{\cdot}42\text{m}$$

$$\text{New draft} = 5{\cdot}25\text{m}$$

Example 2.—A ship floats at a draft of 6·83 m in water of density 1·022 t/m³. What would be her draft in water of density 1·010 t/m³, if her fresh water allowance is 156 mm?

$$\frac{\text{Fa} \times \text{change of density}}{25} = \frac{156\ (1022 - 1010)}{25}$$

$$= 75\text{mm, or } 0{\cdot}08\text{m (sinkage)}$$

$$\text{Old draft} = 6{\cdot}83\text{m}$$

$$\text{New draft} = 6{\cdot}91\text{m}$$

Tonnes per Centimetre Immersion—"T.P.C."—A ship must always displace her own weight of water. If a weight is added to her, it will cause her to sink until she displaces an extra layer of water of equal weight. *The tonnes per centimetre immersion is the weight which must be added to cause the ship to sink one centimetre bodily*—that is, to increase her mean draft by one centimetre. The amount of this will vary with the draft, as we shall see later on.

It is safe to assume that, for all practical purposes, two waterplanes which are one centimetre apart will have the same areas and also that the ship's sides will be vertical between them. The volume of the layer between such waterplanes can thus be considered as having the area of one waterplane and as being one centimetre thick. One centimetre is one-hundredth of a metre so, if A is the area of one waterplane in square metres, the volume of the layer will be:—

$$\frac{A}{100} \text{ cubic metres}$$

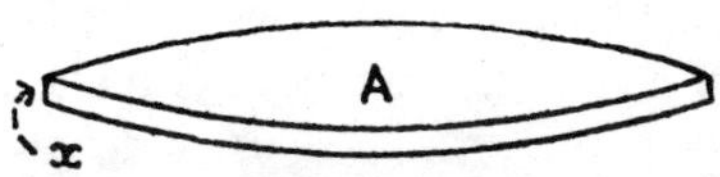

FIG. 45

Fig. 45 illustrates this. The body shewn has flat sides, each having an area, A, and has a depth of x. If we consider the area of the ship's waterplane to be A, and one centimetre to represent x, we can see how the above is arrived at.

In order to submerge the above layer, it is necessary to add to the ship a weight equal to the weight of water which the layer will displace. Since the layer is one centimetre thick, this weight is that which is necessary to sink the ship one centimetre bodily, or the "Tonnes per Centimetre Immersion".

Let δ be the density of the water in which the ship floats.

$$\text{Weight of the layer} = \text{volume} \times \text{density} = \frac{\text{A}}{100} \times \delta$$

$$\text{So, T.P.C.} = \frac{\delta \times \text{A}}{100}$$

Example 1.—A ship floats at a waterplane of area 1520 m². What is her T.P.C. in water of density 1·020 t/m³.

$$\text{T.P.C.} = \frac{\delta \times \text{A}}{100} = \frac{1{\cdot}020 \times 1520}{100} = 15{\cdot}5 \text{ tonnes.}$$

Example 2.—A ship is 120 m long, 18 m beam and the coefficient of fineness of her waterplane is 0·788. Find her T.P.C. in salt water.

$$\text{T.P.C.} = \frac{1{\cdot}025 \times \text{A}}{100} = \frac{1{\cdot}025 \times 120 \times 18 \times 0{\cdot}788}{100}$$

$$= 17{\cdot}4 \text{ tonnes.}$$

To Adjust T.P.C. for Density.—If, as is usual, we are given a ship's T.P.C. in salt water, we can adjust it for water of other densities by direct proportion:—

$$\frac{\text{T.P.C. in density } \delta}{\text{T.P.C. in salt water}} = \frac{\delta}{1{\cdot}025}$$

$$\text{T.P.C. in density } \delta = \text{T.P.C. in salt water x } \frac{\delta}{1{\cdot}025}$$

Example.—A ship has a T.P.C. of 18·35 tonnes in salt water. What would it be in water of density 1·010?

$$\text{T.P.C. in density } 1{\cdot}010 = 18{\cdot}35 \times \frac{1{\cdot}010}{1{\cdot}025}$$

$$= 18{\cdot}08 \text{ tonnes.}$$

Loading to a Given Loadline.—To find out how much to load in order to float at a given loadline on reaching salt water:—

(*a*) Find the ship's present mean draft or freeboard. If she has a list, the freeboards on the Port and Starboard sides will be different: if so, take the mean of the two.

(*b*) Calculate the dock water allowance and apply this to the required salt water draft or freeboard. This will give the allowable draft or freeboard to which the ship can be loaded in the dock water.

(*c*) The difference between (*a*) and (*b*), above, will be the allowable sinkage in the dock water.

(*d*) Adjust the T.P.C. for the density of the dock water.

(*e*) The allowable sinkage, multiplied by the adjusted T.P.C. will be the amount to load to bring the ship to her appropriate load line on reaching salt water.

(*f*) If the ship will use fuel, stores, etc., after leaving her berth, but before reaching salt water, this will reduce her draft to less than that allowable. To compensate for this, extra cargo, equal to the weight of fuel and stores so used, may be loaded before sailing.

Example 1.—A ship is loading in an upriver port, where the density of the water is 1·006 t/m³. Her present freeboards are 1832 mm on the Port side and 1978 mm on the Starboard side. Her statutory summer freeboard is 1856 mm; Fresh water allowance is 148 mm; and her T.P.C. is 18·62 t. On the voyage downriver, she is expected to use 24 tonnes of fuel and 5 tonnes of stores and fresh water. Find how much more cargo she can load to be at her summer load line in salt water.

Present mean freeboard $= \frac{1832 + 1978}{2} = 1905\text{mm}$

Dock water allowance $= \frac{19}{25} \times 148 =$ 112mm

Summer freeboard 1856mm

Allowable freeboard (in dock water) 1744mm

Present ,, ,, ,, ,, 1905mm

Allowable sinkage (in dock water) 161mm (16·1 cm)

T.P.C. in dock water $= \frac{1{\cdot}006}{1{\cdot}025} \times 18{\cdot}62 = 18{\cdot}27\text{t.}$

Total to load = 16·1 × 18·27 294 tonnes

Cargo to replace fuel, etc., used 29 ,,

Cargo to load 323 tonnes

Example 2.—A ship anchors off a Port, in salt water, with the upper edge of her Tropical load line 3·0 cm above water. Her Tropical draft is 7·562 m; T.P.C. is 22·54 t.; F.W.A. is 153 mm. She is to discharge some cargo into lighters in order to enter a dock, where the density of the water is 1·013 t/m³ and the depth on the dock sill is 7·70 m. Find the minimum amount of cargo to discharge in order to cross the dock sill with a clearance of 20 cm, assuming that the vessel remains on an even keel throughout.

Present draft = 7·562 — 0·03 = 7·532m (in salt water)

Required draft in dock water = 7·70 — 0·20 = 7·50m

Dock water allowance $= \frac{12}{25} \times 153 = 73\text{mm}$ 0·073m

Required draft in salt water 7·427m

Present ,, ,, ,, ,, 7·532m

Required rise in salt water 0·105m

(10·5cm)

Cargo to discharge = 10·5 × 22·54 = 237 tonnes

CHAPTER 5

CENTRE OF GRAVITY OF SHIPS

Centre of Gravity of a Ship—"G".—This is often defined as the point through which all the weight of the ship is considered to act vertically downwards.

A ship may be regarded as a hollow shell, inside which weights may be added, removed, or shifted about. Thus, the position of the centre of gravity will change with every condition of loading and must be calculated each time that the ship's stability is to be found. The transverse and longitudinal positions are always considered separately, as in the case of any other body (see Chapter 3). As far as the transverse position is concerned, *G* is usually assumed to be on the centre-line; since if it were not so the ship would list. Longitudinally, it may be forward of, or abaft amidships and is considered accordingly.

"KG".—The vertical height of the centre of gravity above the keel is usually called *"KG"*. This is due to the fact that, in stability diagrams, *K* is usually taken to denote the keel and *G* the centre of gravity.

"Light KG."—The height of *G* above the keel in the light ship, before any cargo, stores or fuel are placed on board, is calculated by Naval Architects. It is given to the seaman in the ship's stability information.

Before a ship is built, the *KG* is estimated, usually by comparison with some existing ship of similar size and lines, although in some unusual cases it is actually calculated approximately. The *KG* of the completed ship, when light, can be found by means of the "Inclining Experiment", which will be described later.

Shift of "G".—The centre of gravity of a ship obeys the same laws as that of any other body. Let us summarise the conclusions which we drew in Chapter 3 with regard to this matter.

G moves directly towards the centre of gravity of any weight added to the ship, directly away from the centre of gravity of any weight taken away from the ship and parallel to the shift of the centre of gravity of any weight moved from one place to another.

The distance through which G will move can be found from the formula:—

$$GG_1 = \frac{w \times d}{W}$$

Where $GG_1 =$ the shift of G.
$w =$ the weight shifted
$d =$ (for weights removed or added)—the distance from G to the centre of gravity of the weight.
(for weights shifted)—the distance through which the centre of gravity of the weight is moved.
$W =$ (for added weights)—the displacement after the weight has been added.
(for weights removed)—the displacement after the weight has been removed.
(for weights shifted)—the displacement of the ship.

Example 1.—Find the shift of the centre of gravity of a ship of 7000 tonnes displacement, if a weight of 50 tonnes is shifted for a distance of 80 metres.

$$GG_1 = \frac{w \times d}{W} = \frac{50 \times 80}{7000} = 0{\cdot}57 \text{ metres}$$

Example 2.—A ship has a displacement of 3200 tonnes. What would be the shift of her centre of gravity if a weight of 200 tonnes is added at a distance of 60 metres from the original position of her centre of gravity?

$$GG_1 = \frac{w \times d}{W}$$ (Where W equals the displacement after adding the 200 tonnes)

$$GG_1 = \frac{200 \times 60}{3400} = 3{\cdot}53 \text{ metres}$$

Example 3.—A weight of 48 tonnes is removed from a lighter, the centre of gravity of this weight being 2·0 metres above the keel. What will be the new KG of the lighter if its original displacement and KG were 690 tonnes and 6·2 metres, respectively?

$$d = 6{\cdot}2 - 2{\cdot}0 = 4{\cdot}2 \text{ metres}$$
$$W = \text{Displacement after weight has been removed}$$
$$= 690 - 48 \ldots\ldots 642 \text{ tonnes}$$
$$GG_1 = \frac{w \times d}{W} = \frac{48 \times 4{\cdot}2}{642} = 0{\cdot}31 \text{ metre}$$

G moves vertically upwards, away from the weight removed, so:—

$$\text{New } KG = \text{Old } KG + GG_1 = 6{\cdot}2 + 0{\cdot}3$$
$$= 6{\cdot}5 \text{ metres}$$

"KG" for any Condition of Loading.—The Naval Architects who build a ship, give the seaman her light KG and displacement. They cannot give him the KG for any other condition of loading, so he must find this for himself if he requires it.

When weights are added to the ship, G will move upwards or downwards according to whether the centre of gravity of the weight is above or below that of the ship. (Note that we are here only considering the shift of G in the vertical direction.)

In this case, we could use the method just described to find successive shifts of G due to each weight: but this would be laborious and errors could easily creep into the calculations. A better and more simple method of finding the new KG is to take moments about a horizontal line through the keel, known as the "base line".

Now, moment is weight multiplied by length of lever. In this case, the length of lever will be the distance from the base line to the centre of gravity of the weight (sometimes written as Kg).

So the moment of each weight $= w \times Kg$

And final $KG =$ Total moment $\div$ total weight.

The method now becomes:—

(*a*) The ship's original displacement and KG are multiplied together to give her original moment.

(*b*) Each weight, added or removed, is multiplied by its height above the base line, to give its moment. Added weights and moments are added to those of the ship. Weights removed and their moments are subtracted.

(*c*) The total moment, divided by the total weight, will give the new KG of the ship.

Example 1.—A ship arrives in port with a displacement of 4250 tonnes and KG of 5·96 metres. She then loads 520 tonnes at 6·3 m above the keel; 1250 tonnes at 4·2 m above the keel; 810 tonnes at 11·6 m above the keel. She also discharges 605 tonnes from 2·4 m above the keel.
What will then be her KG?

	Weight	Height	Moment
Ship	4250	5·96	25330
Loaded	520	6·3	3276
,,	1250	4·2	5250
,,	810	11·6	9396
	6830		43252
Discharged	— 605	2·4	— 1452
Totals	6225		41800

$$\text{New } KG = \frac{\text{Total moment}}{\text{Total weight}} = \frac{18400}{6225} = 6{\cdot}71 \text{ metres}$$

Example 2.—A ship has a light displacement of 4559 t and light *KG* of 6·62 m. Find her *KG* and displacement when she has the following weights on board:—

Cargo	730 tonnes at	5·1	metres above the keel.
,,	480 ,, ,,	9·3	,, ,, ,, ,,
,,	1470 ,, ,,	4·7	,, ,, ,, ,,
,,	860 ,, ,,	8·5	,, ,, ,, ,,
Fuel	355 ,, ,,	0·6	,, ,, ,, ,,
Stores	60 ,, ,,	12·4	,, ,, ,, ,,
F. Water	80 ,, ,,	8·8	,, ,, ,, ,,

	Weight	Height	Moment
Light Ship	4559	6·62	30181
Cargo	730	5·1	3723
,,	480	9·3	4464
,,	1470	4·7	6909
,,	860	8·5	7310
Fuel	355	0·6	213
Stores	60	12·4	744
F. Water	80	8·8	704
Totals	8594		54248

New displacement = 8594 tonnes.

$$\text{New } KG = \frac{\text{Total moment}}{\text{Total weight}} = \frac{54248}{8594} = 6{\cdot}31 \text{ metres.}$$

NOTE.—The *KG*s which we have found above are often called the "Solid *KG*". If the ship has free liquid on board, it may be necessary to apply a correction to this, in order to find the "Fluid *KG*", as we shall see in Chapter 9.

Deadweight Moment.—This term is sometimes used to describe the sum of the moments about the base line of all the items which make up the ship's deadweight (i.e. cargo, fuel, stores, etc.). It does not include the moment of the light ship. So we can say that:—

Total moment = Moment of light ship + Deadweight moment.

For instance, in Example 2, above:—

Moment of light ship	30181t/m
Deadweight moment	24067t/m
Total moment	54248t/m

Real and Virtual Centres of Gravity.—If a weight is free to move about, its effect is as if its centre of gravity were somewhere above the weight, instead of being at its real centre of gravity. This imaginary point is called the "Virtual Centre of Gravity" and for stability calculations it is regarded as being the centre of gravity of the weight.

An example of this can be found in the case of a weight which is being lifted by a derrick. Suppose the weight to be one centimetre off the bottom of the hold. The real centre of gravity has scarcely moved—certainly not enough to affect the centre of gravity of the ship appreciably. If the weight is now free to swing, however, a new (virtual) centre of gravity appears at the head of the derrick, so that G will move as if the centre of gravity of the weight were actually at the head of the derrick and not down in the hold.

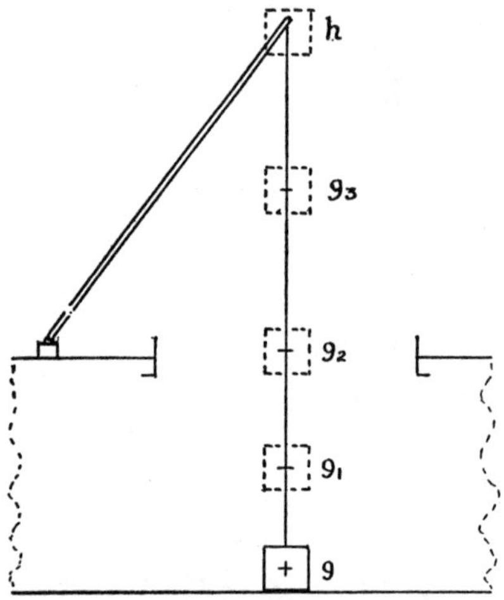

FIG 46

Fig. 45 shews a weight being lifted by a derrick. When the weight is resting on the bottom of the hold, its centre of gravity and its effect on the ship is at g. As soon as the weight has been lifted clear of the bottom of the hold, so that it is free to swing, its virtual centre of gravity appears at the head of the derrick, h. The effect on the ship is now as if the weight were actually moved from g to h. The virtual centre of gravity will remain at h, whatever the actual height of the weight (i.e. whether it be at g_1, g_2, or g_3) provided that the weight is free to swing.

A similar effect occurs in "slack" tanks, which contain liquid which is free to move. This is called "Free Surface Effect" and causes a virtual centre of gravity to appear at some height above the actual centre of gravity of the liquid in the tank. This effect is discussed more fully in Chapter 9.

A somewhat similar effect is also found where chilled meat is hung from beams in a hold and is more or less free to move.

Example 1.—A ship displaces 6200 t and has a KG of 6·12 m when a lift of 80 t is resting on the inner bottom, at 2·50 m above the keel. What would be the ship's KG when the lift was being hoisted by a derrick, the head of which is 21 m above the keel.

Head of derrick is $21 - 2{\cdot}5 = 18{\cdot}5$ m above c.g. of lift. So c.g. of the lift rises 18·5 m when lifting.

$$GG_1 = \frac{w \times d}{W} = \frac{80 \times 18{\cdot}5}{6200} = 0{\cdot}24\text{m (upwards)}$$

$$\text{Old } KG = 6{\cdot}12\text{m}$$

$$\text{New } KG = 6{\cdot}36\text{m}$$

Example 2.—A ship displaces 7400 t and her centre of gravity is on the centre line at 6·50 m above the keel. A locomotive, weighing 100 t, is to be lifted from the quay by means of a derrick. The head of the derrick, when lifting, will be 25 m above the keel and 12 m to starboard of the centre line. Find the position of *G* whilst the locomotive is being lifted.

To find the new *KG*:—

	Weight	Height	Moment
Ship	7400	6·50	48100
	100	25	2500
Loco.	7500		50600

$$\text{New } KG = \frac{50600}{7500} = 6{\cdot}75\text{m.}$$

To find the horizontal shift of *G*:—

$$GG_1 = \frac{w \times d}{W} = \frac{100 \times 12}{7500} = 0{\cdot}16\text{m to starboard of c/l.}$$

The Effect of Tanks on "G".—When a tank is filled with water or oil, weight is added to the ship and *G* will move directly towards the centre of gravity of the tank. When a tank is emptied, the reverse happens and *G* moves directly away from the centre of gravity of the tank. The distance it moves will be the same as for any other weight added or taken away and can be found by the same formula, viz.:—

$$GG_1 = \frac{w \times d}{W}$$

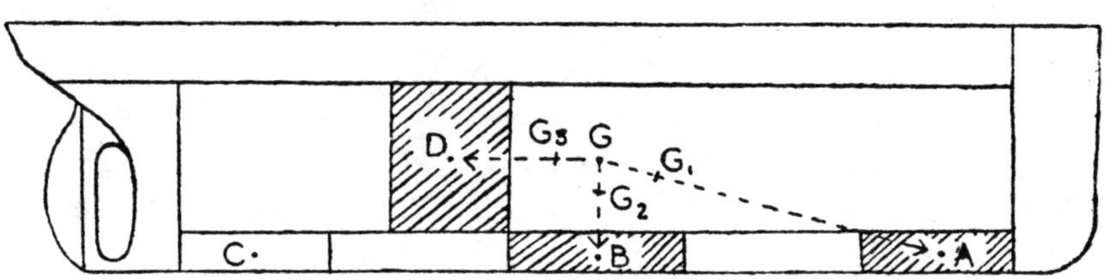

FIG. 47A

One or two examples will illustrate this more fully.

If a forward double-bottom tank, having its centre of gravity at *A* were filled, *G* would move forward and downward to G_1. If a double-bottom tank, having its centre of gravity at B were filled, G would move vertically downward to G_2. If a tank aft, at C, were filled, G would move directly toward C, downward and aft.

In the case of a deep tank, the one shown in Fig. 47 (*a*) is abaft G and has its centre of gravity (D) level with it, so in this case, G will move horizontally aft to G_3, directly towards D. The relative positions of the centre of gravity of a deep tank and that of the ship vary considerably with different ships and different conditions of loading. So the above case is only an example and not an invariable rule. Filling a deep tank may have a very different effect from that shown here and the only rule that can be laid down is that G will move directly toward the centre of gravity of the tank.

When tanks are emptied, G will move in exactly the opposite direction from the above in each case. That is, if tank A were emptied, it would cause G to move upwards and aft; tank B would cause it to move vertically upward; tank C, upward and forward; and tank D, directly forward.

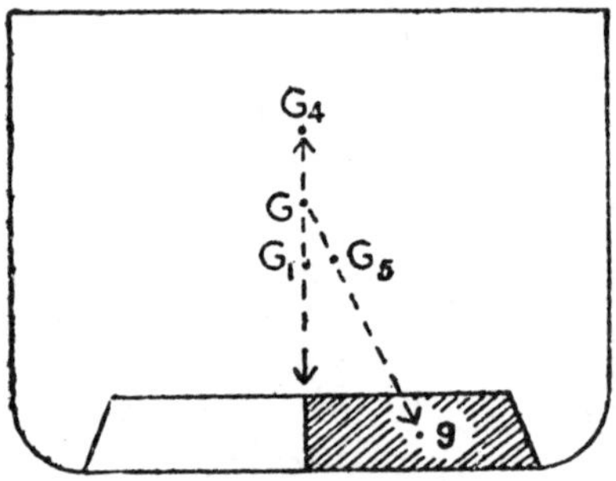

FIG. 47B

Now consider the transverse effect of filling the same tanks. As long as the tanks are symmetrical about the centre-line of the ship, their centres of gravity will be on that line. Thus, so far as the transverse stability of the ship is concerned, G is merely assumed to move vertically upwards or downward. For instance, if we filled or emptied the whole tank here, G would move down to G_1 or up to G_4.

If the tank were not symmetrical about the centre-line of the ship, its centre of gravity would be to one side of that line, so that G would move sideways as well as up or down. For instance, if we only filled one side of a double-bottom tank, with its centre of gravity at g, then G would move in the direction shown to G_5.

When it is necessary to calculate the effect of filling or emptying tanks they are treated as ordinary weights added or removed. The vertical shift of G is found by taking the weight and KG of the ship, weight and KG of the liquid in the tank and then adding or subtracting the moments (see "*KG* for any condition of loading").

It must be remembered that, in all the above cases, the tanks are considered as being completely filled or emptied. If they are not, the free surface of the liquid may cause a virtual centre of gravity, which will produce some very different effects, as described in Chapter 9.

CHAPTER 6

CENTRES OF BUOYANCY AND FLOTATION

Centre of Buoyancy—"B".—This is the geometrical centre of the underwater part of the ship. That is, it is the centre of gravity of the water which has been displaced.

The force of buoyancy is considered to act vertically upward through the centre of buoyancy, with a force equal to the weight of the displaced water. Since, by the Law of Archimedes, a floating body displaces its own weight of water, this force must be equal to the weight of the ship.

The transverse and longitudinal positions of B are always considered separately. In the transverse direction, B will always be on the centre-line as long as the ship is upright, but will move out to one side of this line when she heels. Longitudinally, B may be a little forward of or abaft the centre line.*

"KB".—This is the vertical height of the centre of buoyancy above the keel. For box-shaped ships, floating upright, KB is always equal to one-half of the draft; this is obvious if we consider B as the centre of gravity of the underwater body. For ship shapes, KB is usually between 0·55 and 0·60 of the draft and can be found by seamen from the ship's stability curves or scales.

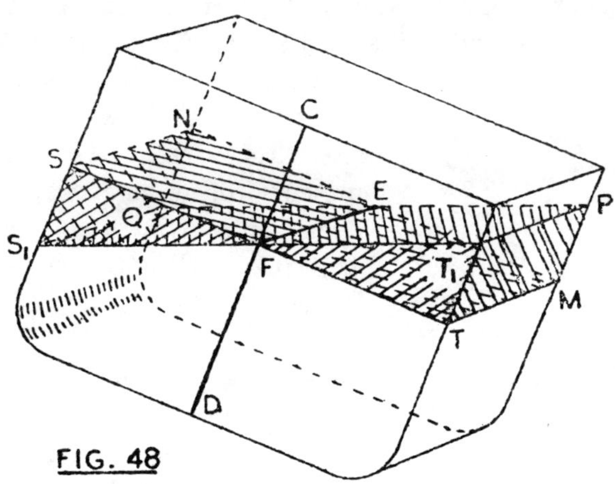

FIG. 48

Centre of Flotation — "F".—This is the point about which the ship heels and trims. Consider a ship which is heeled to a small angle, as shown in Fig. 48. Let the plane $STMN$ be the original waterplane and the plane S_1T_1PQ be the new waterplane. The wedge SS_1FNQE has emerged from the water, whilst the wedge FTT_1PME has become immersed. These are known as the "emerged wedge" and "immersed wedge", respectively.

By the Law of Archimedes, a ship must displace her own weight of water at all times if she is to remain afloat. Thus, the volume which she displaces

---

* The method of finding the position of B is given in Chapter 3.

when heeled must be the same as that which she displaced when upright; so that the volumes of the immersed wedge and of the emerged wedge must be equal. When the sides of the ship are parallel, the line forming the apex of each wedge must divide each waterplane into exactly equal areas. For instance, in the figure, the line EF must be such that the area $SNEF$ is equal to the area $TMEF$ and the area S_1QEF is equal to the area T_1PEF. This will hold good whether the ship swings longitudinally or transversely, or, for that matter, in any direction. It is obvious that all such "centre-lines" must cut each other at one point—the geometrical centre of each waterplane; or, in other words, its centre of gravity.

In box-shaped ships, the centres of gravity of the upright and heeled waterplanes must coincide, unless the deck-edge becomes submerged, or the bilge emerges from the water. In the case of ship-shapes this is not strictly true, but for small angles of heel or trim it can be taken as correct for all practical purposes. This gives us a new definition for the centre of flotation, namely that *the centre of flotation is the centre of gravity of a ship's waterplane.*

The transverse position of the centre of flotation is always at the centre line of the waterplane; that is, the intersection of the waterplane and the centre-line of the ship. Longitudinally, it is in the waterplane and at the centre-line for box shapes; but may be a little abaft or forward of the centre-line in ship shapes. Chapter 3 shews how it may be found.

Shift of "B".—The centre of buoyancy has been defined as the centre of gravity of the water which has been displaced by a ship. It may, therefore, be expected to obey the same laws as any other centre of gravity. Fig. 49 represents the ship shown n Fig. 48, as it would be seen transversely. ST represents the original waterplane and S_1T_1 the new waterplane when the vessel is heeled. SS_1F is the emerged wedge and TT_1F the immersed wedge, which have their centres of gravity at g and g_1, respectively. B is the position of the centre of buoyancy before the ship heeled. The effect of heeling the ship is the same as if we took the emerged wedge away and placed it in the position of the immersed wedge. Since the centre of gravity of a body moves parallel to the shift of the centre of gravity

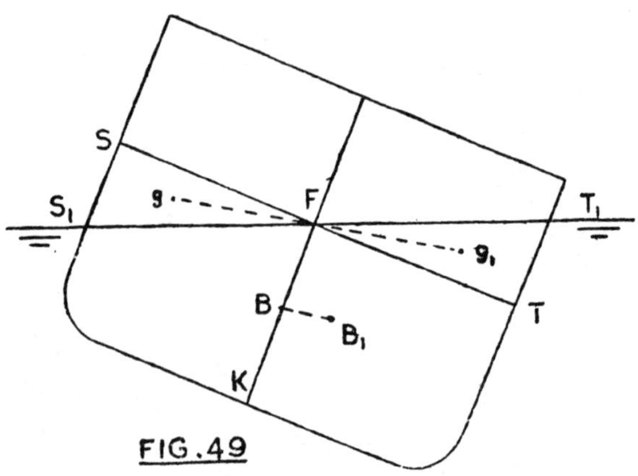

FIG. 49

of any weight shifted, B must move out to B_1, in a direction parallel to the line gg_1

The distance which B will shift can be found by taking moments, in the same way as for the shift of any centre of gravity. Let W be the displacement of the ship and w the weight of water in either wedge.

Now, for the centre of gravity, $W \times GG_1 = w \times d$

So, for the centre of buoyancy, $W \times BB_1 = w \times gg_1$

The weight of any body equals its volume in cubic metres multiplied by its density. So if V represents the volume of displacement of the ship; v, the volume of either wedge; and δ the density of the water; the above formula can be modified, thus:—

$$W \times BB_1 = w \times gg_1$$

$$BB_1 = \frac{w \times gg_1}{W}$$

$$BB_1 = \frac{v \times \delta \times gg_1}{V \times \delta}$$

$$BB_1 = \frac{v \times gg_1}{V}$$

All the above will also be true for longitudinal inclinations of the ship.

Example.—A ship displaces 28,800 cubic metres and is heeled so that the volume of the immersed wedge is 1550 cubic metres. The distance between the centres of gravity of the immersed and emerged wedges is found to be 9 metres. Find the shift of B.

$$BB_1 = \frac{v \times gg_1}{V} = \frac{1550 \times 9}{28800} = 0{\cdot}484 \text{ metres}$$

Horizontal and Vertical Components of the Shift of "B".—For more advanced problems in stability, it is sometimes necessary to know how far B will shift in either the horizontal or the vertical direction only, when the ship heels. Fig. 50 shows the same ship as in the last example, heeled and with B moved out to B_1. In the new vertical direction, B has moved downwards for the distance RB_1. In the

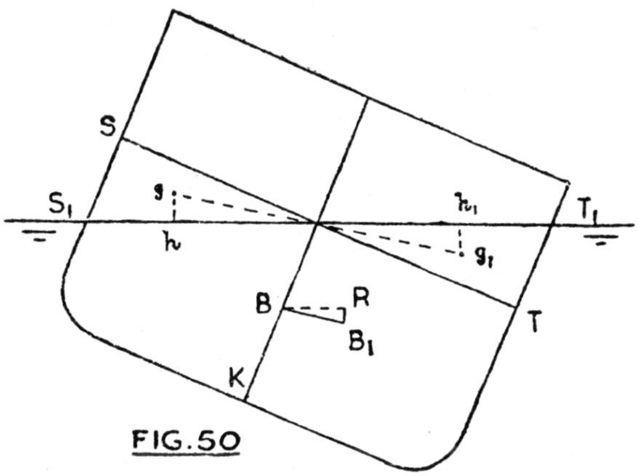

FIG. 50

new horizontal direction, it has moved out for the distance BR. These two components of BB_1 can be found as follows:—

Drop perpendiculars from g and g_1 on to the new waterplane S_1T_1. Let these perpendiculars be gh and g_1h_1.

For the horizontal shift, BR:—The shift of g in the horizontal direction is hh_1. If we take moments in this direction about any vertical line, the moment of the wedges must equal that of the ship. Thus, using volume in lieu of weight, as in the last example.

$$V \times BR = v \times hh_1$$

$$BR = \frac{v \times hh_1}{V}$$

For the vertical shift, B_1R:—g has risen for a vertical distance of gh above the horizontal, S_1T_1. g_1 has fallen for a vertical distance of g_1h_1 below the horizontal. Thus, the total vertical shift of g is $(gh + g_1h_1)$. If we take moments about any horizontal line, we shall see that:—

$$V \times B_1R = v\,(gh + g_1h_1)$$

$$B_1R = \frac{v\,(gh + g_1h\ )}{V}$$

CHAPTER 7

THE RIGHTING LEVER AND METACENTRE

Equilibrium of Ships.—We have seen in Chapter 3 that a body's state of equilibrium determines whether, when it is tilted, it will right itself, remain as it is, or turn over. Seamen are, naturally, very much concerned as to whether their ships will remain upright and so the study of equilibrium forms an important part of ship stability.

In the normal ship, the centre of gravity is always higher than the centre of buoyancy; that is, KG is greater than KB. The force of gravity acts vertically downwards through the former and the force of buoyancy vertically upwards through the latter. As we have already seen, these two forces must be equal. It has been shewn in Chapter 3 that the equilibrium of a tilted body depends on the relative positions of the centre of gravity and the point of support. Unless one is vertically over the other, the body will try to turn in one direction or the other. This will hold good for ships, if we substitute "centre of buoyancy" for "point of support"; thus for a ship to remain at rest, G must be vertically over B.

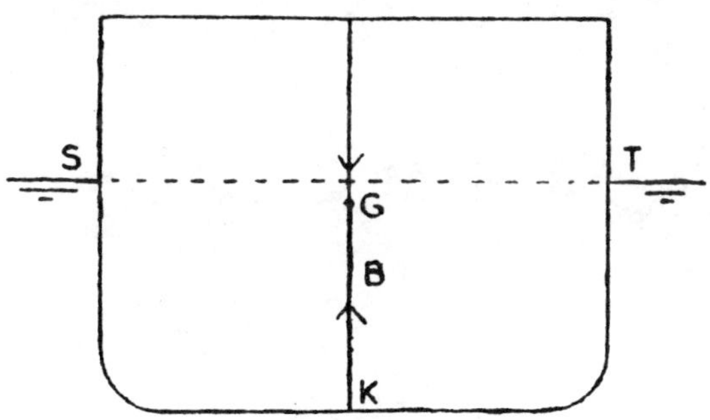

FIG. 51

When a ship is upright and correctly loaded, B and G will both be on the centre-line, as shewn in Fig. 51. The forces of gravity and buoyancy will be equal and opposite and there will be no tendency for the ship to move from the upright. As soon as any other force acts on her, however, she will heel. The centre of buoyancy will move out towards the low side and the results of this movement will decide whether the ship is in stable, neutral or unstable equilibrium.

In the next six sections we shall consider Figs. 52 (*a*), (*b*) and (*c*), which show a ship inclined by some external force such as wave action. In each case, B has moved out to B_1, but since no weights have been moved on board, G remains in its original position on the centre line. The force of gravity now

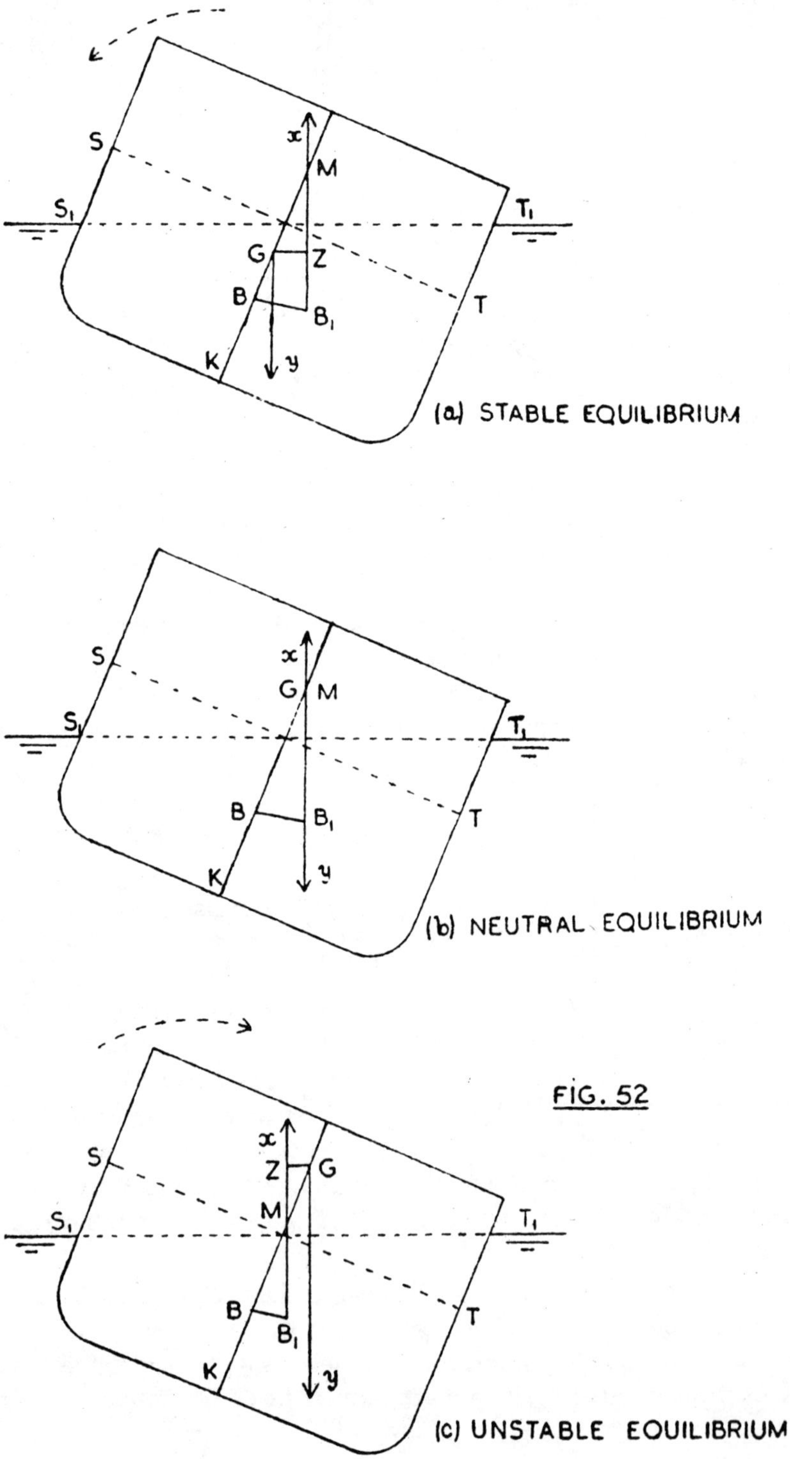

(a) STABLE EQUILIBRIUM

(b) NEUTRAL EQUILIBRIUM

(c) UNSTABLE EQUILIBRIUM

FIG. 52

acts vertically downwards through G, in the direction Gy and the force of buoyancy vertically upwards through the new centre of buoyancy, in the direction B_1x. These two forces form a "couple", and, in the case of Figs. 52 (*a*) and (*c*), are trying to turn the ship in one direction or the other.

The Righting Lever—"GZ".—In Figs. 52 (*a*) and (*c*), a horizontal line, GZ, has been drawn perpendicular to B_1x. This perpendicular distance between G and the direction of the action of the force of buoyancy is called the "righting lever", on the ends of which the forces of gravity and buoyancy act to produce a turning movement. It can be seen that when this lever is on the immersed side of the ship, she will try to right herself (Fig. 52*a*); when it is on the other side, she will try to heel further over (Fig. 52*c*); when it does not exist, there will be no turning effect (Fig. 52*b*).

The Metacentre—"M".—In Fig. 52, the point at which the force of buoyancy, acting in the direction B_1x, cuts the centre-line ot the ship, has been marked M. For small angles of heel, up to about 10 or 15 degrees, the shift of the centre of buoyancy is very small and M, which is caused by this shift, may be regarded as a fixed point. This point is called "the metacentre".

For larger angles of heel, B moves out more quickly and this causes M to move upward, so that it can no longer be regarded as a fixed point. A point called the "pro-metacentre" is sometimes considered to exist in this case. This point changes its position with every change of the angle of heel and may not even be on the centre-line, so it is not used for ordinary stability calculations.

Metacentric Height—"GM".—This is the distance between the centre of gravity and the metacentre. It has a definite relationship with GZ and, since it is easier to find than the latter, it is often used for calculating stability at small angles of heel. It cannot be used for angles of over about 15 degrees, since the metacentre then ceases to exist. GM is termed "positive" if G is below M and "negative" if G is above M.

NOTE.—The term "Height of the metacentre" is sometimes used in stability to denote the height of M above the keel; that is, KM. It should not be confused with metacentric height, or GM.

Stable Equilibrium.—A ship is said to be in stable equilibrium when, if she were inclined by some external force, she would try to return to the upright. If we consider Fig. 52 (*a*), we shall see that this condition will exist when:—

1. For small angles of heel, the ship has a positive GM.
2. For any angle of heel, the righting lever, GZ, is on the low side of the ship.

Neutral Equilibrium.—A ship which, if heeled by some external force, would have no tendency, either to return to the upright, or to heel further over, is said to be in neutral equilibrium. We can see from Fig. 52 (*b*) that this will occur when G and M coincide, so that there is no GM or GZ.

Unstable Equilibrium.—We say that a ship is in unstable equilibrium when, if inclined by some external force, she would try to heel still further. Fig. 52 (*c*) shews that this will occur when:—

1. For small angles of heel, the ship has a negative GM.
2. For any angle of heel, the righting lever, GZ, is on the high side of the ship.

It should be remembered that a ship in this condition will not necessarily capsize. As she heels further over, the centre of buoyancy will usually move further outwards and may become vertically under the centre of gravity at some larger angle of heel. She will then have developed neutral equilibrium at that angle.

Longitudinal Metacentric Height—"GM_L".—We have so far only considered transverse stability in this chapter. A ship also has a longitudinal metacentre and metacentric height, which obey the same rules as the transverse ones, although the longitudinal metacentre is not in the same position. To distinguish them, the longitudinal metacentre and metacentric height are usually denoted by M_L and GM_L. We shall consider them later, as they are used in calculating trim and for several other purposes.

The longitudinal righting lever is not considered in stability, although it obviously exists. The angles of inclination in this direction are so small that GM_L, which is more easily found, can be used for all necessary calculations.

CHAPTER 8

TRANSVERSE STATICAL STABILITY

Moment of Statical Stability.—This is the moment which will try to return a ship to the upright when she is heeled. It is often termed "positive" if it tends to right the ship and "negative" if it tries to cause her to heel still further over.

We have seen in Chapter 3 that moment is equal to force multiplied by the length of the lever. In the case of statical stability, the lever is GZ and the force acting on this lever is equal to the weight (i.e. displacement) of the ship. So, if W be the displacement and GZ the righting lever:—

$$\text{Moment of Statical Stability} = W \times GZ.$$

For any given condition of loading, when the displacement is constant, a ship's moment of statical stability will increase or decrease, or be positive or negative, with GZ. So we can say that, whilst the measure of statical stability is $W \times GZ$, the length and direction of GZ alone is an indication of the ship's statical stability at any angle of heel.

Relation Between "GM" and "GZ".—There is a definite relationship between GZ and GM and the latter increases, decreases and becomes positive or negative with GZ.

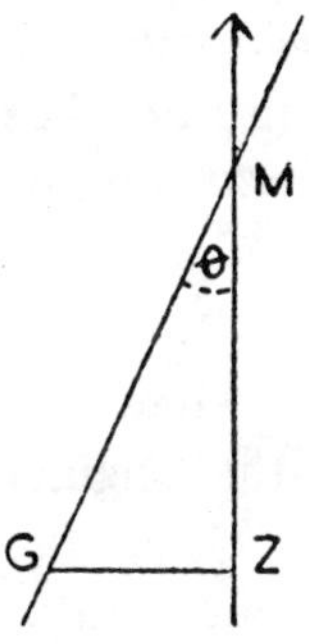

FIG. 53

Fig. 53 represents the GZM triangle from Fig. 52 (*a*). The angle Z is a right angle, so that, if θ is the angle of heel:—

$$\frac{GZ}{GM} = \sin\theta$$

$$GZ = GM \times \sin\theta$$

This means that for angles of heel of less than 10° or 15° we can use GM as the indication of statical stability, instead of GZ. This is an advantage, since the former is more easily found, but it must be remembered that M is not considered to exist for larger angles, so that we must use GZ for these latter.

Initial Stability and Range of Stability.—Initial stability is the statical stability of a ship at a very small angle of heel, and is considered to be indicated by GM. It determines whether the ship will be "stiff" or "tender" and if she is likely to develop a list during a voyage. It gives no real indication as to how the ship will behave at angles of heel for which M is not considered to exist.

Range of stability is the angular range over which a ship will have positive statical stability. It is important because it indicates the angle to which the ship could heel before she would capsize.

A ship's initial stability does not necessarily indicate what her range of stability is likely to be, or vice versa. The two have little to connect them and a ship with a large intial stability may have either a large or small range of stability. It is also quite possible for a ship to have negative initial stability, yet to become stable at a small angle of heel and thereafter to be able to heel to quite a large angle before she capsizes.

Calculation of a Ship's Stability.—When a ship is built, the naval architects calculate her displacement, deadweight and the height of the centre of buoyancy above the keel (*KB*). They also find the distance of the metacentre above the centre of buoyancy (*BM*) and, by adding this to the *KB*, obtain the height of the metacentre (*KM*).

Once the ship is nearly completed, the "Inclining Experiment" is performed to find the metacentric height (*GM*) of the ship in the light condition. This is subtracted from the light *KM* to give the height of the centre of gravity above the keel (the light *KG*).

The above information is tabulated in the "Deadweight Scale", or given in the form of graphs called "Curves of Stability". The righting levers for various angles of heel and for assumed *KG*'s are also calculated and added to the stability information; usually in the form of "Cross Curves". Care is taken to see that the range of stability is adequate to ensure the safety of the ship at any reasonable angle of heel if she is properly loaded.

This completes the naval architects' part of the work. Armed with the above information, the seaman can calculate the *KG* of his ship at any stage of loading, and can thus find her *KM* and *GM*, her righting levers at various angles of heel and her approximate range of stability.

Calculation of BM for Ship Shapes.—Let a ship be heeled by some external force, as shewn in Fig. 54. *ST* is the original waterline and S_1T_1 is the new one. The original centre of buoyancy, *B*, has moved out to B_1. *g* and g_1 are the centres of gravity of the emerged and immersed wedges, SCS_1 and TCT_1, respectively. The Metacentre (*M*) has appeared.

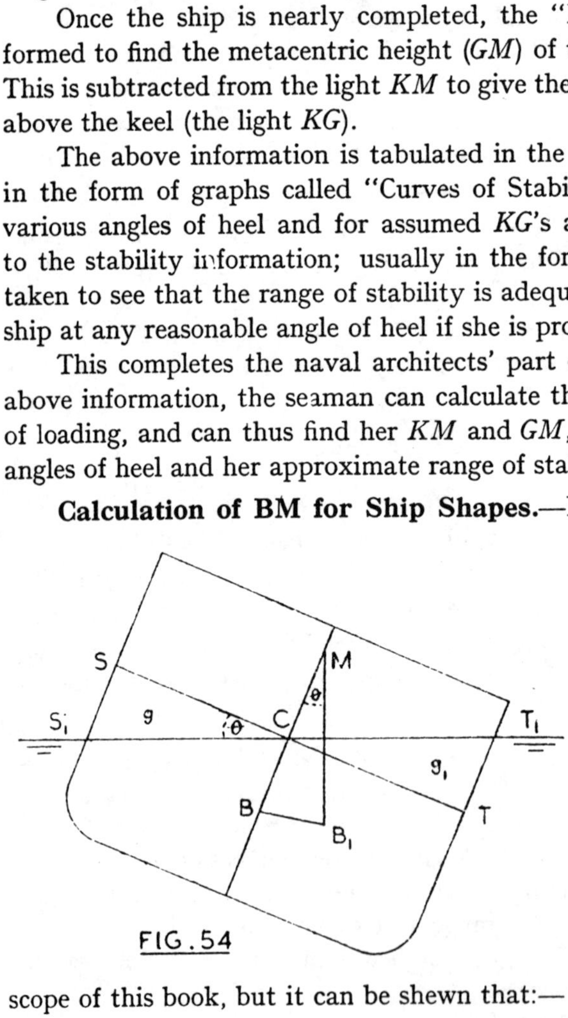

FIG. 54

The proof of the following formula is outside the scope of this book, but it can be shewn that:—

Where I = Moment of inertia of the waterplane
V = Ship's volume of displacement

$$BM = \frac{I}{V}$$

Example.—The inclining experiment is performed on a ship and her GM is found to be 1·90 metres. Her displacement in salt water is 3200 tonnes and the moment of inertia of her waterplane was 17,070. If her KB is 1·62 metres, what is her KG?

$$\text{Volume of displacement} = \frac{3200}{1 \cdot 025} = 3122 \text{ m}^3$$

$$BM = \frac{I}{V} = \frac{17070}{3122} = 5 \cdot 47 \text{ metres}$$

$$\begin{aligned} BM &= 5 \cdot 47 \text{ metres} \\ KB &= 1 \cdot 62 \quad \text{,,} \\ \hline KM &= 7 \cdot 09 \quad \text{,,} \\ GM &= 1 \cdot 90 \quad \text{,,} \\ \hline KG &= 5 \cdot 19 \quad \text{,,} \end{aligned}$$

Calculation of BM for Box Shapes.—A box-shaped vessel has a rectangular waterplane. It was shewn, in Chapter 3, that the moment of inertia of a rectangle about its centre line is

$$\frac{lb^3}{12}$$

where l is the length and b the breadth of the waterplane.

We can thus substitute this for I in the last formula:—

$$BM = \frac{I}{V}$$

$$BM = \frac{lb^3}{12V}$$

When D equals the draft, $V = l \times b \times D$

So, $$BM = \frac{lb^3}{12 \times l \times b \times D}$$

$$BM = \frac{b^2}{12D}$$

It must be remembered that *the above formula holds good for box shapes only.*

Example.—A box-shaped vessel is 60 metres long, 10 metres broad and floats at a draft of 3·0 metres. Find her *BM* and height of the metacentre.

$$BM = \frac{b^2}{12D} = \frac{10^2}{12 \times 3} = 2{\cdot}78 \text{ metres}$$

$$KB = \tfrac{1}{2} \text{ draft} = 1{\cdot}50 \text{ metres}$$

$$BM = 2{\cdot}78 \quad ,,$$

$$KM = 4{\cdot}28 \quad ,,$$

Approximate Formula for BM.—A close approximation for *BM*, which is sometimes useful, can be found by the following formula:—

Where b = the ship's breadth.
D = her mean draft.
a = a coefficient.

$$BM = \frac{ab^2}{D}$$

a is about 0·08 in very fine ships and about 0·10 in very full-formed ships. Its average value for merchant ships is about 0·09.

The Inclining Experiment.—This is performed to find the ship's light *GM* and hence her light *KG*. It consists of shifting weights transversely across the deck of a ship when the latter is free to heel. The angle of heel is measured by the shift of a plumb-bob along a batten.

Certain conditions are necessary for this experiment, if it is to give good results, viz:—

(*a*) Mooring lines must be slack and the ship clear of the wharf, so that she may heel freely.

(*b*) The water must be smooth and there should be little or no wind. If there is any wind, the ship should be head-on or stern-on to it.

(*c*) There must be no free surface of water in the ship. The bilges must be dry and boilers and tanks dry or pressed up.

(*d*) All moveable weights must be properly secured.

(*e*) All persons should be ashore, except the men actually engaged in the experiment.

(*f*) The ship must be upright at the beginning of the experiment.

When this experiment is performed in practice, four weights are generally used, two on each side of the ship. These are shifted alternately, first one and then both, across the deck. Two or three plumb-lines are used and all weights and plumb-lines are identical in order that they may provide a reliable check on each other.

For the purpose of proving the formula, etc., the effect of shifting one weight and of using one plumb-line only is considered.

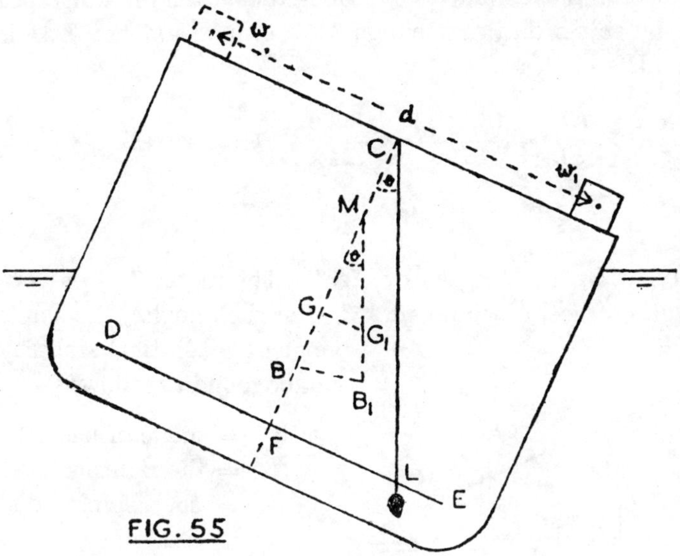

FIG. 55

In the figure, *DE* represents a batten, fixed horizontally across the ship and having the point, *F*, at which it cuts the centre-line, marked on it. *CL* is a plumb-line, suspended at *C* and free to move across the batten when the ship heels.

The weight, w, is shifted across the deck to w_1, through a distance of d metres. G then moves out to G_1 and the ship heels until B has moved to B_1, vertically under both G_1 and M. The plumb-line moves out across the batten for the distance FL.

Let θ be the angle of heel and let W be the ship's displacement.

Consider the shift of G, as described in Chapter 4:—

$$GG_1 = \frac{w \times d}{W} \qquad (1)$$

Consider the triangle FCL:—

The angle F is a right angle and the angle C is equal to θ.

$$\text{So,} \quad \frac{CF}{FL} = \cot\theta \qquad (2)$$

Consider the triangle MGG_1:—

w has been shifted across the deck at right angles to the centre-line. Since G moves parallel to the shift of the weight. the angle G must be a right angle. The angle M is equal to θ.

$$\text{So,} \quad \frac{GM}{GG_1} = \cot\theta$$

$$GM = GG_1 \times \cot\theta$$

substituting for GG_1 and for $\cot\theta$, from formulae (1) and (2), above, this gives us:—

$$GM = \frac{w \times d}{W} \times \frac{CF}{FL}$$

Example.—A weight of 15 tonnes is moved horizontally across the deck of a ship for a distance of 11·0 metres. The ship heels so that a pendulum, suspended 8·0 metres above a horizontal batten, moves out along the batten for a distance of 244 millimetres. The ship's displacement is 3150 tonnes and her KM is 6·12 metres. Find the GM and KG.

$$GM = \frac{w \times d}{W} \times \frac{CF}{FL} = \frac{15 \times 11}{3150} \times \frac{8000}{244} = 1{\cdot}72 \text{ metres}$$

$$KG = KM - GM = 6{\cdot}12 - 1{\cdot}72 = 4{\cdot}40 \text{ metres}$$

Statical Stability at Small Angles of Heel.—The metacentre is considered to exist for angles of heel up to about 15°. For such angles, the ship's moment of statical stability can be found as follows:—

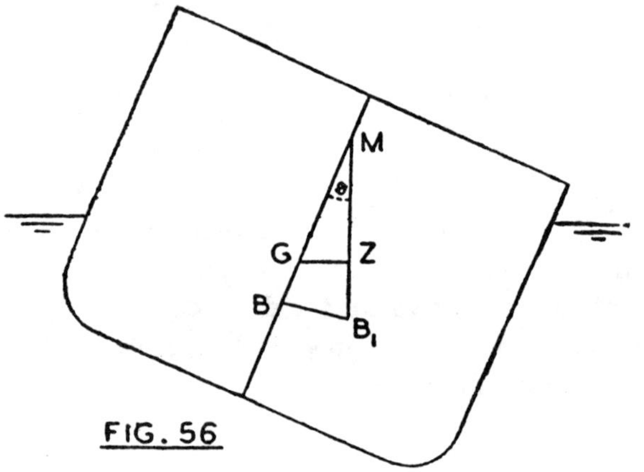

FIG. 56

θ = angle of heel
GZ = the righting lever
W = the ship's displacement

We have already seen that, since the angle GZM is a right angle,

$$GZ = GM \times \sin\theta$$

Also that, moment of statical stability $= W \times GZ$

From the above it can be seen that, for small angles of heel:—

$$\text{Moment of statical stability} = W \times GZ$$
$$\text{Moment of statical stability} = W \times GM \times \sin\theta$$

Example.—A ship of 6240 tonnes displacement has a GM of 0·67 metre. Find her moment of statical stability at an angle of heel of 9°.

$$\begin{aligned}\text{Moment of statical stability} &= W \times GM \times \sin\theta \\ &= 6240 \times 0{\cdot}67 \times \sin 9° \\ &= 654 \text{ tonne-metres}\end{aligned}$$

Statical Stability at any Angle of Heel.—The following is known as "Atwood's Formula", and holds good for any angle of heel.

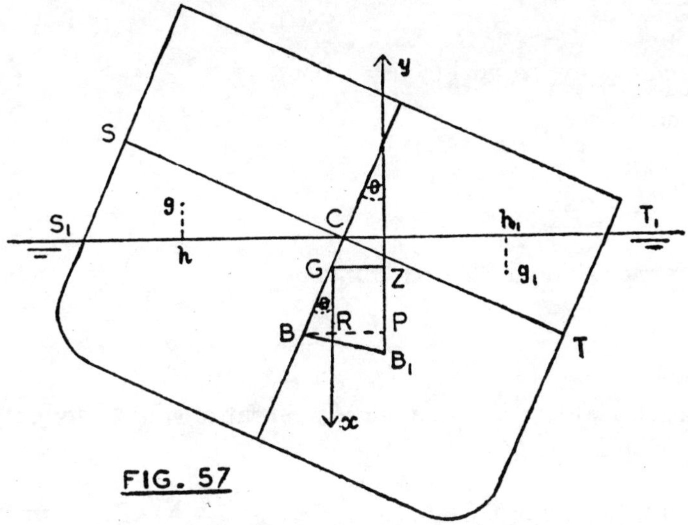

FIG. 57

Let θ = Angle of heel.
v = Volume of the immersed or emerged wedge.
V = Volume of displacement of the ship.
W = Displacement of the ship, in tonnes.
GZ = Righting lever.
ST = Original waterline.
S_1T_1 = New waterline.
SCS_1 = The emerged wedge.
TCT_1 = The immersed wedge.
g and g_1 = The centres of gravity of the wedges.

When the ship heels, B moves out to B_1. Let Gx and B_1y be the vertical direction of the actions of the forces of gravity and buoyancy, respectively.

From g and g_1, drop perpendiculars on to S_1T_1 and let these be gh and g_1h_1. From B draw BP parallel to S_1T_1 and let it cut Gx at R.

hh_1 is the horizontal component of the shift of g.

BP is the horizontal component of the shift of B.

It has been shown in Chapter 6 (shift of B), that in such a case:—

$$BP = \frac{v \times hh_1}{V} \quad \ldots \quad (1)$$

Consider the triangle GBR:—

The angle G is equal to θ and the angle R is a right angle, so,

$$\frac{BR}{BG} = \sin\theta$$

$$BR = BG \times \sin\theta \quad \ldots \quad (2)$$

To find GZ:—

GX and B_1y are parallel (construction)

BP and GZ ,, ,,

$$\text{So, } GZ = RP$$

$$= BP - BR$$

Substituting for BP and BR from formulae (1) and (2), this becomes:—

$$GZ = \frac{v \times hh_1}{V} - BG \times \sin\theta$$

Alternatively we can say that, since moment of statical stability is equal to $W \times GZ$, then:—

$$\text{Moment of Statical Stability} = W\left\{\frac{v \times hh_1}{V} - BG \times \sin\theta\right\}$$

"GZ" by the "Wall-Sided Formula".—This formula is not as accurate as Atwood's, but is sometimes used to find GZ when the positions of G, B and M are known. It states that:—

$$GZ = \sin\theta\,(GM + \tfrac{1}{2}BM \times \tan^2\theta)$$

It is based on the assumption that the ship is "Wall-Sided": that is, that her sides are vertical at all points along the waterplane, when she is upright. Although this assumption is not strictly true, the formula will give reasonably accurate results for angles of heel of up to 20° to 30° in ships of normal shape, provided that the deck edge has not become submerged.

Loll, or List.—A ship may develop a list for one of two reasons:—

(*a*) If the centre of gravity is out of the centre-line of the ship.

(*b*) If the ship has a negative GM.

These conditions are usually caused by faulty loading of the cargo and are generally avoidable if the weights in the ship are properly distributed. They are not necessarily dangerous, provided that the ship has an adequate range of stability, but are obviously bad seamanship. In either case the ship will heel over until she is in neutral equilibrium; that is, until B has moved out sufficiently to come vertically under G.

The first condition can occur in either stiff or tender ships and the list will always be towards that side of the centre-line to which G has moved. In the second condition the list may be to either side and may, under the influence of external forces, change from one side to the other. It may also increase or decrease if weights are taken away from, or added to the ship.

Angle of Heel due to "G" being out of the Centre-line.—If a weight is shifted or loaded so that G comes out of the ship's centre-line, the ship must heel until G comes vertically over B. This is precisely what happens in the "inclining experiment", as we have already seen. In that case, the heel was caused by weights shifted. Now let us consider how weights may be loaded so as to cause or to correct heel.

Case 1.—Consider a ship which is upright to begin with, but in which a weight of w tonnes is then loaded at a distance of d metres off the centre line, as shewn in Fig. 57 (a). If the ship's centre of gravity was originally at G_0, it will now move upwards and outwards to G_1. $G_0 G$ will be the, vertical component of this shift and GG_1 the horizontal component. The ship's new metacentric height will be GM.

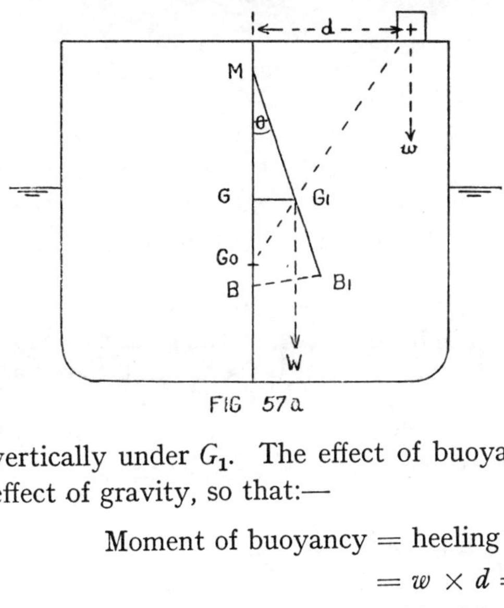

FIG 57a

The moment trying to heel the ship is $w \times d$ t/m, which equals $W \times GG_1$ t/m.

The ship will now heel until, at some angle θ, the centre of buoyancy has moved out to come vertically under G_1. The effect of buoyancy will then exactly counteract the effect of gravity, so that:—

$$\text{Moment of buoyancy} = \text{heeling moment}$$
$$= w \times d = W \times GG_1$$

But in the triangle $GG_1 M$: $GG_1 = GM \times \tan\theta$

So; $w \times d = W \times GM \times \tan\theta$

(Where W and GM are the new displacement and metacentric height, respectively.)

If several weights are loaded or discharged, we must take their combined moments in lieu of $w \times d$ in the above formula.

Case 2.—Now take the case of a ship which is already heeled because of uneven distribution of weights on board, as shewn in Fig. 57 (*b*). Let us see how to bring her upright. The ship's centre of gravity must be somewhere off her centre line, say at G_1. So the heel must be caused by the ship's weight, W, acting vertically downwards through G_1. Let GG_1 be the perpendicular distance of G_1 from the ship's centre line: then the downward force through G_1, at right angles to GG_1, will be $W \times \cos\theta$.

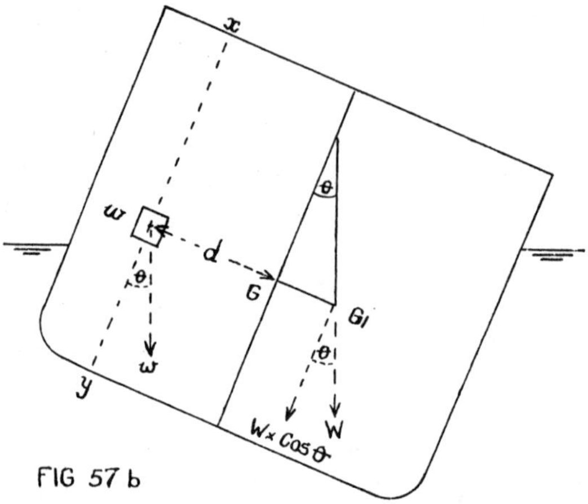

FIG 57 b

So, the moment heeling the ship $= W \times GG_1 \times \cos\theta$.

Now, to bring the ship upright, let us load a weight of w tonnes at a distance of d metres from the ship's centre line, so as to balance the heeling moment.

The moment to right the ship $= w \times d \times \cos\theta$

And, $w \times d \times \cos\theta = W \times GG_1 \times \cos\theta$

Cos θ cancels out, so:— $w \times d = W \times GG_1$

But, $GG_1 = GM \times \tan\theta$

So, $w \times d = W \times GM \times \tan\theta$

Which is the same formula as in the last case, except that we are here using the original (heeled) W and GM..

Note that, in the figure, the weight, w, is shewn at the same height as G. This is done for simplicity: but in fact, the height at which the weight is loaded does not matter in this case, provided that it does not cause the ship to become unstable. In other words, we could load the same weight anywhere on the line xy, in the figure, to bring the ship upright.

General Rules.—The following will apply to both the above cases.

(*a*) $w \times d = W \times GM \times \tan\theta$;

(*b*) Always use the **heeled** W and GM, in the above;

(*c*) If a number of weights are loaded or discharged, use their combined moments for $w \times d$.

(*d*) When a ship, which is heeled, is to be brought upright, the height at which the weight is loaded is immaterial, provided that it does not render the ship unstable.

Example 1.—A ship, which is upright, has a GM of 2·64 m and displacement of 3200 t. A weight of 160 t is then loaded into the wing of a tween deck, so that it is 4·25 m above G and 6·36 m off the ship's centre line. Find the angle to which the ship will heel

In this case, the heeled W and GM will be those after the weight has been loaded.

$$G_0 G = \frac{w \times d}{W} = \frac{160 \times 4.25}{3200 + 160} = 0{\cdot}20\text{m}$$

$$\text{New } GM = \text{old } GM - G_0 G = 2{\cdot}64 - 0{\cdot}20 = 2{\cdot}44\text{m}$$

$$w \times d = W \times GM \times \tan\theta$$

$$160 \times 6{\cdot}36 = 3360 \times 2{\cdot}44 \times \tan\theta$$

$$\text{Tan } \theta = \frac{160 \times 6{\cdot}36}{3360 \times 2{\cdot}44} = 0{\cdot}1241$$

$$\theta = 7{\cdot}1^\circ$$

Example 2.—A ship displaces 5700 t and is listed 6° to starboard. Her KG is 6·24 m and her KM is 7·01 m. Calculate the weight of cargo to load into the port wing of a hold, at a distance of 12·0 m off the centre line, in order to bring the ship upright.

In this case, the heeled W and GM will be the original ones.

$$w \times d = W \times GM \times \tan\theta$$

$$w \times 12{\cdot}0 = 5700 \times 0{\cdot}77 \times \tan 6^\circ$$

$$w = \frac{5700 \times 0{\cdot}77 \times 0{\cdot}1051}{12{\cdot}0} = 38{\cdot}4 \text{ tonnes}$$

Example 3.—A ship displaces 8820 t, has a KG of 6·73 m and is upright. She then loads:—

540 tonnes at 3·0m above the keel and on the centre line
300 ,, ,, 9·0m ,, ,, ,, ,, 2·5m to port of c/l
200 ,, ,, 5·0m ,, ,, ,, ,, 6·8m to port of c/l
280 ,, ,, 7·5m ,, ,, ,, ,, 4·5m to starboard of c/l

If the KM is then 7·81m, find the heel, if any.

Weight	Height	Vert. Moment	Dist.	Hor. Moment
8820	6·73	59359	0	0
540	3·0	1620	0	0
300	9·0	2700	2·5	750 (to port)
200	5·0	1000	6·8	1360 (to port)
280	7·5	2100	4·5	1260 (to starboard)
10140		66779		850 (to port)

$$\text{New } KG = \frac{66779}{10140} = 6{\cdot}59\text{m}$$

$$\text{New } GM = KM - KG = 7{\cdot}81 - 6{\cdot}59 = 1{\cdot}22\text{m}$$

$$\text{Heeling moment (for } w \times d) = W \times GM \times \tan\theta$$

$$850 = 10140 \times 1{\cdot}22 \times \tan\theta$$

$$\text{Tan } \theta = \frac{850}{10140 \times 1{\cdot}22} = 0{\cdot}0687$$

$$\text{Heel} = 3{\cdot}9^\circ \text{ to port}$$

Example 4.—A ship displaces 6470 t, has a KG of 5·97 m, KM of 7·62 m and is heeled 5° to starboard. 400 tonnes of cargo are to be loaded into the wings of a tween deck at distances of 5·0 m off the centre line to port and 7·0 m off the centre line to starboard. Find the weight to load in each wing in order that the ship may then be upright.

Load w tonnes into the port wing: $(400—w)$ tonnes into the starboard wing.

Moment to port = 5w t/m

Moment to starboard = $7(400 - w)$ t/m

The ship is heeled to starboard, so we require the greater moment in the port wing in order to bring her upright.

So, moment required to right ship = $5w - 7(400 - w)$ t/m

$$\text{Righting moment} = W \times GM \times \tan\theta$$

$$5w - 7(400 - w) = 6470 \times 1{\cdot}65 \times \tan 5°$$

$$5w - 2800 + 7w = 6470 \times 1{\cdot}65 \times 0{\cdot}0875$$

$$12w - 2800 = 934{\cdot}1$$

$$w = \frac{934{\cdot}1 + 2800}{12} = 311 \text{ tonnes}$$

So we must load 311 tonnes in the port wing and 89 tonnes in the starboard wing.

Angle of Loll due to a Negative "GM".—It was stated in the last chapter, that a ship may develop a list through having a negative GM. In such a case, she will heel until she is in neutral equilibrium: that is, until B has moved out to a position where it is vertically under G.

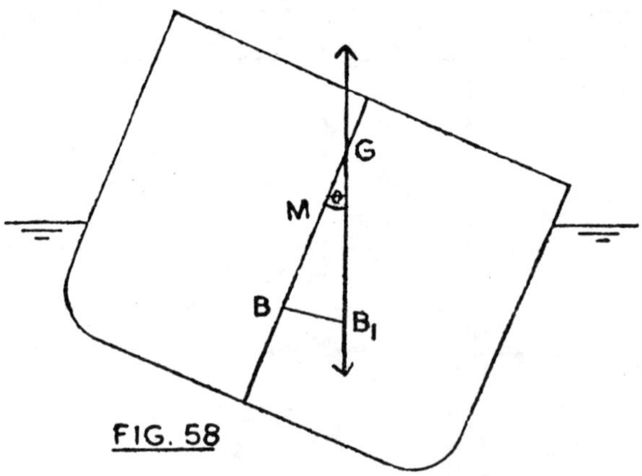

FIG. 58

The figure represents a ship in the above condition. Two things are obvious:—

1. That GM must be negative in the first instance

2. That, after the ship has heeled and B is vertically under G, there can be no GZ.

We have seen, from the wall-sided formula, that:—

$$GZ = \sin\theta\,(GM + \tfrac{1}{2}BM \times \tan^2\theta)$$

But it has just been shewn that $GZ = O$, so:—

$$O = \sin\theta\,(GM + \tfrac{1}{2}BM \times \tan^2\theta)$$

For two quantities, when multiplied together, to be equal to O, one at least of them must be O. This means that either,

(1.) $\sin\theta = O$:

or (2.) $GM + \frac{1}{2}BM \times \tan^2\theta = O$

If $\sin\theta$ equals O, then the angle θ must also equal O, and the ship would be upright. This is obviously impossible, since the ship has a list, so we can conclude that:—

$$GM + \tfrac{1}{2}BM \times \tan^2\theta = O$$

Since GM is negative, we get from this, that:—

$$-GM + \tfrac{1}{2}BM \times \tan^2\theta = O$$

$$\tfrac{1}{2}BM \times \tan^2\theta = GM$$

$$\tan^2\theta = \frac{2GM}{BM}$$

$$\tan\theta = \sqrt{\frac{2GM}{BM}}$$

Example.—A ship has a displacement of 4500 tonnes and a GM of 0·24 metres. A weight of 270 tonnes is then loaded on deck, so that its centre of gravity is 5·0 metres above that of the ship and the BM is then found to be 4·40 metres. Find the new GM and the angle of loll, if any.

New displacement $= 4500 + 270 = 4770$ tonnes

$$GG_1 = \frac{w \times d}{W} = \frac{270 \times 5{\cdot}0}{4770} = 0{\cdot}28 \text{ metres}$$

New $GM = 0{\cdot}24 - 0{\cdot}28 = -0{\cdot}04$ metres

To find the angle of loll:—

$$\tan\theta = \sqrt{\frac{2GM}{BM}} = \sqrt{\frac{2 \times 0{\cdot}04}{4{\cdot}40}} = 0{\cdot}1348$$

Angle of loll $= 7\frac{3}{4}°$

CHAPTER 9

FREE SURFACE EFFECT

The Effect of Free Surface of Liquids.—If a tank is completely filled with liquid, the latter becomes, in effect, a solid mass. It can be treated in exactly the same way as any other weight in the ship; that is, its weight can be regarded as being concentrated at its actual centre of gravity.

In a tank which is only partly filled, the surface of the liquid is free to move and possesses inertia. The moment of inertia of this free surface about its own centre-line causes a virtual centre of gravity to appear at some height above it. The effect on the ship's stability will then be as if a weight, equal to the weight of the liquid in the tank, were raised from its position in the tank to the position of the virtual centre of gravity.

Fig. 59 shows a ship which is heeled and which has free water in a double-bottom tank.

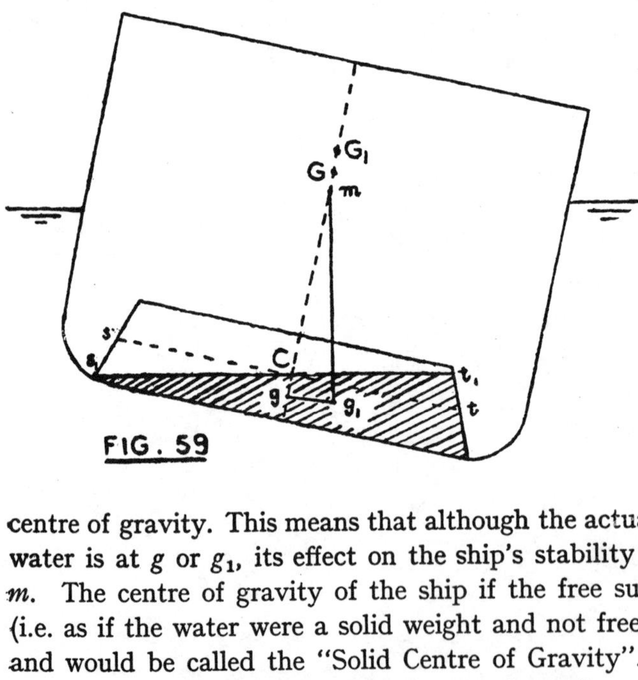

FIG. 59

st is the original surface of the water and *g* its original centre of gravity. When the ship heels, its surface becomes s_1t_1 and its centre of gravity shifts out to g_1. g_1m is the new vertical through g_1. The water now behaves almost as if it were a weight, free to swing and hanging from *m*, which is its virtual centre of gravity. This means that although the actual centre of gravity of the water is at *g* or g_1, its effect on the ship's stability is as if it were placed at *m*. The centre of gravity of the ship if the free surface effect were ignored (i.e. as if the water were a solid weight and not free to move) would be at *G* and would be called the "Solid Centre of Gravity". When the free surface is taken into account, the ship's centre of gravity would be at G_1 and would be called the "Fluid Centre of Gravity".

When the GM of a ship is being calculated, the fluid KG and GM should always be taken as the true values, if free surface of liquid exists.

We have already seen that, for a ship, when I is the moment of inertia of the waterplane and V the volume of displacement:—

$$BM = \frac{I}{V}$$

If we consider Fig. 59, it is apparent that scs_1 and tct_1 are equivalent to the immersed and emerged wedges, respectively, of a ship: that the shift of g, out to g_1 is parallel to the shift of the centre of gravity of these wedges: that m occurs at the intersection of the centre-line and the vertical line through g_1.

This is obviously similar to what happens when B moves out to produce M, in the case of a ship.

So if i is the moment of inertia of the free surface and v the volume of water in the tank:—

$$gm = \frac{i}{v} \quad \text{. . . . . (1)}$$

The effect of the virtual centre of gravity is the same as would be that of a solid weight placed at m: that is, as if the weight of the water in the tank were shifted from g to m.

So, if W is the displacement of the ship and w the weight of water in the tank:—

$$GG_1 = \frac{w \times gm}{W} \quad \text{(Shift of "G"—Chapter 5)}$$

Let v be the volume of water in the tank, V the volume of displacement of the ship, δ_1 the density of the liquid in the tank and δ the density of the water in which the ship floats. Then:—

$$w = v \times \delta_1$$

$$W = V \times \delta$$

So,

$$GG_1 = \frac{v \times \delta_1 \times gm}{V \times \delta}$$

If we now substitute for gm, from formula (1):—

$$GG_1 = \frac{v \times \delta_1 \times i}{V \times \delta \times v}$$

$$GG_1 = \frac{i}{V} \times \frac{\delta_1}{\delta}$$

Example.—A ship displaces 9750 t and has a solid GM of 0·96 m when floating in water of density 1·020 t/m³. Free surface exists in a tank which contains water of density 1·005 t/m³. The moment of inertia of this free surface is 1640. What is the fluid GM?

$$V = W \div \delta = 9750 \div 1{\cdot}020$$

$$\text{Loss of } GM = \frac{i}{V} \times \frac{\delta_1}{\delta} = \frac{1640 \times 1{\cdot}005}{9750 \div 1{\cdot}020 \times 1{\cdot}020}$$

$$= 0{\cdot}17\text{m}$$

Solid GM 0·96m

Fluid GM 0·79m

The Effect of Rectangular Free Surfaces.—The free surface in a tank is often a rectangular area. In this case, the calculation of its effect is much simplified. If l is the length and b the breadth of such a surface, its moment of inertia will be found by the formula:—

$$i = \frac{lb^3}{12}$$

Since

$$GG_1 = \frac{i}{V} \times \frac{\delta_1}{\delta}$$

$$GG_1 = \frac{lb^3}{12V} \times \frac{\delta_1}{\delta}$$

Free Surface Effect when Tanks are Filled or Emptied.—When liquid is run into, or removed from a tank so as to cause free surface, it will affect the ship's stability in two ways:—

(*a*) The solid KG and displacement will be changed by the weight added or removed.

(*b*) There will be a rise of G caused by the free surface. This will have to be added to the new solid KG to give the fluid KG.

Example.—A ship displaces 5154 t, floats in sea water and has a KG of 6·10 m when a double bottom tank is empty. The tank is rectangular, 15 m long, 12 m wide and 1·3 m deep. What would be the ship's GM after fresh water has been run into the tank to a depth of 0·80 m, if the KM is then 7·45 m.

Added volume = 15 × 12 × 0·8 = 144 m³ (144 t of fresh water).

	Weight	Height	Moment
Ship	5154	6·10	31439
Water	144	0·40	58
	5298		31497

$$\text{Solid } KG = \frac{31497}{5298} = 5{\cdot}95\text{m}$$

$$\text{Rise of } G \text{ due to free surface} = \frac{lb^3}{12V} \times \frac{\delta_1}{\delta}$$

$$= \frac{15 \times 12^3}{12 \times 5298 \div 1{\cdot}025} \times \frac{1{\cdot}000}{1{\cdot}025} = 0{\cdot}41\text{m}$$

$$\text{Solid } KG = 5{\cdot}95\text{m}$$

$$\text{Fluid } KG = 6{\cdot}36\text{m}$$

$$KM = 7{\cdot}45\text{m}$$

$$\text{Fluid } GM = 1{\cdot}09\text{m}$$

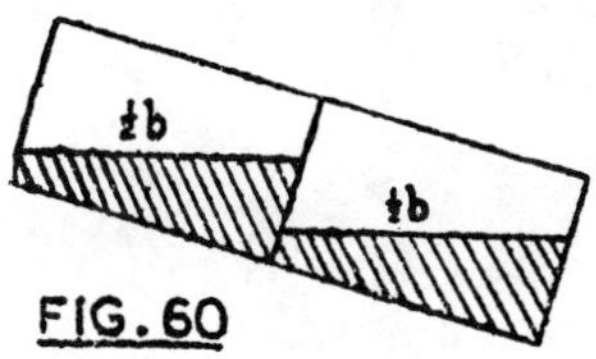

FIG. 60

Free Surface in Divided Tanks.—Figure 60 shows a tank having a rectangular free surface and divided at the centre-line. Let l be the length and b the breadth of the tank. The breadth of the free surface on either side of the tank will thus be $\frac{1}{2}b$.

$$\text{Rise of } G \text{ due to free surface on one side} = \frac{l \times (\frac{1}{2}b)^3}{12V}$$

$$= \frac{l}{12V} \times \frac{b^3}{8}$$

GG_1 is due to the free surface on two sides, however, so:—

$$GG_1 = 2 \times \frac{l}{12V} \times \frac{b^3}{8}$$

$$= \frac{2}{8} \times \frac{lb^3}{12V}$$

So

$$GG_1 = \frac{1}{4} \times \frac{lb^3}{12V}$$

If we compare this with the formula found in the last section, we shall see that it is one quarter of the latter. So, in the case of a rectangular free surface, dividing the tank at the centre-line will decrease the rise of G to one quarter of what it would be in an undivided tank. Since that rise is only due to free surface, it does not matter whether the tank is divided by a watertight bulkhead or only by a washplate, as long as the latter extends to below the surface. In the case of surfaces other than rectangular ones, the decrease may not be the same, but it will be considerable in any case.

It can be proved, in the same way, that dividing the tank into three would decrease the rise of G due to free surface to approximately one-ninth of that for an undivided tank. Dividing the tank into four parts will decrease the free surface effect to about one-sixteenth of the above.

If free surface exists in more than one tank, this will cause a total decrease in the ship's GM equal to the sum of the effects for each tank.

Example 1.—A ship is floating in salt water and displaces 6000 t, has a solid KG of 5·810 m, and KM of 6·424 m. Free surface exists in a rectangular bunker tank, 3·0 m long and 12·0 m wide, divided at the centre line and containing oil of density 0·890 t/m³. What is the ship' sfluid GM?

$$\text{Free surface effect} = \frac{lb^3}{12V} \times \frac{\delta_1}{\delta} \times \frac{1}{4}$$

$$= \frac{3 \times 12^3}{12 \times 6000 \div 1{\cdot}025} \times \frac{0{\cdot}890}{1{\cdot}025} \times \frac{1}{4}$$

$$= 0{\cdot}016\text{m}$$

$$\text{Solid } KG = 5{\cdot}810\text{m}$$

$$\text{Fluid } KG = 5{\cdot}826\text{m}$$

$$KM = 6{\cdot}424\text{m}$$

$$\text{Fluid } GM = 0{\cdot}598\text{m}$$

Example 2.—A ship, floating in salt water, displaces 8440 t, has KG of 6·21 m and KM of 7·21 m, when a double bottom tank is full of fresh water. The tank is rectangular, 18 m long, 20 m wide, 1·5 m deep and has a centre girder and two side girders, equally spaced. Find the ship's GM after 360 t of fresh water have been pumped out of the tank, leaving it slack?

$$\text{Depth of water removed} = \frac{360}{18 \times 20} = 1{\cdot}00\text{m}$$

c.g. of water removed = 0·5m from top of tank

Kg of water removed = 1·5 — 0·5 = 1·00m

	Weight	Height	Moment
Ship	8440	6·21	52412
Water	— 360	1·0	— 360
	8080		52052

$$\text{Solid } KG = \frac{52052}{8080} = 6{\cdot}442\text{m}$$

The surface is divided into 4 parts

So, $$\text{Free surface effect} = \frac{lb^3}{12V} \times \frac{\delta_1}{\delta_1} \times \frac{1}{16}$$

$$= \frac{18 \times 20^3}{12 \times 8080 \div 1{\cdot}025} \times \frac{1{\cdot}000}{1{\cdot}025} \times \frac{1}{16}$$

$$= 0{\cdot}09\text{m}$$

$$\text{Solid } KG = 6{\cdot}44\text{m}$$

$$\text{Fluid } KG = 6{\cdot}53\text{m}$$

$$KM = 7{\cdot}21\text{m}$$

$$\text{Fluid } GM \quad 0{\cdot}68\text{m}$$

Free Surface Moments.—One of the requirements of the Load Line Rules is that a ship shall be provided with a statement shewing the loss of *GM* due to free surface in each tank. The Department of Trade now recommend that this be given as "Free Surface Moments", assuming that the liquid in the tank has a density of 1·000 t/m.

Where i is the moment of inertia of the free surface (allowing for longitudinal subdivisions):—

$$\text{Free surface moment} = i \times 1{\cdot}025 \times \left(\frac{1{\cdot}000}{1{\cdot}025}\right)$$

Which amounts to saying that, numerically:—

$$\text{Free surface moment} = i$$

Now, $$\text{loss of } GM \text{ due to free surface} = \frac{i}{V} \times \frac{\delta_1}{\delta}$$

But $V = W \div \delta$

So, $$\text{loss of } GM = \frac{i}{W \div \delta} \times \frac{\delta_1}{\delta} = \frac{i \times \delta_1}{W}$$

$$= \text{Free surface moment} \times \frac{\delta_1}{W}$$

Example.—A ship has a solid *GM* of 1·02 m and displaces 7600 t when a tank is partly filled with oil of relative density 0·850. If the free surface moment of this tank is 1250 t/m, find the fluid *GM*?

$$\text{Loss of } GM = \text{F.S.M.} \times \frac{\delta_1}{W} = 1250 \times \frac{0{\cdot}850}{7600}$$

$$= 0{\cdot}14\text{m}$$

$$\text{Solid } GM = 1{\cdot}02\text{ m}$$

$$\text{Fluid } GM = 0{\cdot}88\text{m}$$

CHAPTER 10

TRANSVERSE STATICAL STABILITY IN PRACTICE

Factors Affecting Statical Stability.—Statical stability is governed principally by:—

(*a*) The position of the ship's centre of gravity.

(*b*) The form of the ship.

The position of the centre of gravity depends on the loading of the cargo and other weights in the ship. It affects the statical stability, because it is one of the factors which determine the length of the righting lever, *GZ*.

The form of the ship decides the shape of the emerged and immersed wedges when the vessel heels. These in their turn will determine the shift of the centre of buoyancy and hence the length of *GZ*; or alternatively, the position of *M* and hence the *GM*.

An example will best show the effect of the above. Let us consider a graph showing a ship's moment of statical stability at various angles of heel.

Curve *A* is for a vessel 160 metres long, 20 metres beam, 8 metres draft, 3 metres freeboard and having a *KG* of 7·00 metres. The maximum righting

FIG. 61

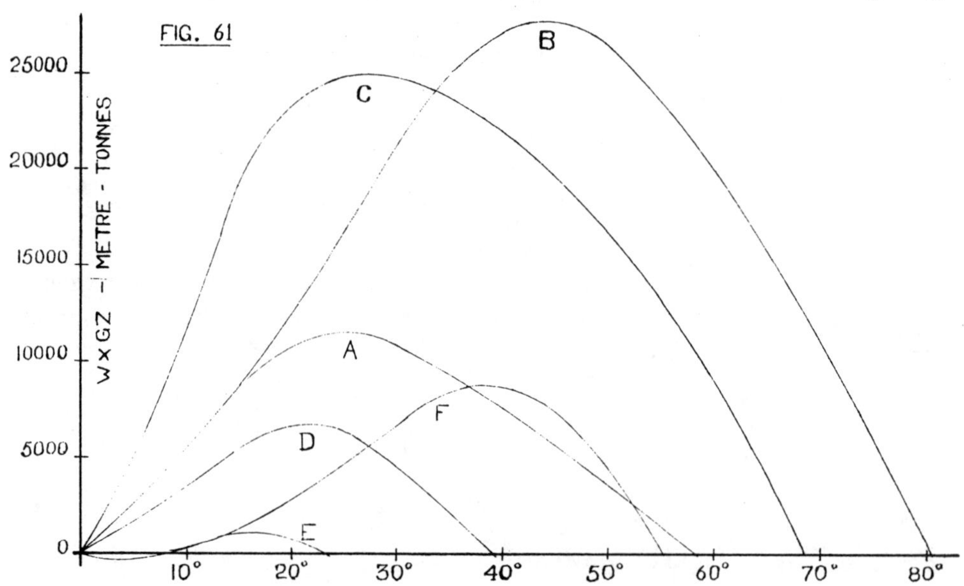

moment for this ship is about 11,600 tonne-metres and occurs at 23° of heel. Her range of stability is 58°.

Curve *B* shews the effect of adding 2 metres of freeboard to the above ship, if all other details remain the same. The two curves run together at first, but curve *B* continues to rise to a maximum of about 27,500 tonne-metres at 48° of heel. The range of stability has increased to 81°.

Curve *C* shews the effect of adding 2 metres of beam to the original ship in curve *A*. The maximum stability has increased to 25,000 tonne-metres, but it only occurs at about 25° heel. The range of stability has increased to 68°.

The effect of raising the centre of gravity of ship *A* by 0·5 metre is shewn in curve *D*. The maximum stability is now nearly 7000 tonne-metres and the range is 39° — a considerable reduction in each case.

Curve *E* shews the effect of raising *G*, in ship *A*, by 1·20 metres, so as to give her a negative *GM* of 0·03 metres. The negative stability causes her to loll to an angle of about 7°, but thereafter she develops positive stability and has a range to 23°.

Curve *F* is an example of what would happen if ship *A* had *G* raised by 1·20 metres (as in curve *E*), but at the same time had the freeboard increased by 2 metres. In this case she will still loll to 7°; but thereafter she will have a range to about 55°, because of the increased freeboard.

Let us tabulate these results:—

Ship	*GM* metres	Maximum stability		Range of stability
		Tonne-metres	Angle of heel	
A	1·17	11,600	23°	58°
B	1·17	27,500	43°	81°
C	2·04	24,700	25°	66°
D	0·67	6,800	22°	39°
E	—0·03	1,100	17°	23°
F	—0·03	8,800	39°	55°

From the above we can draw the following conclusions:—

(*a*) Increase of freeboard does not affect initial stability, but increases range of stability.

(*b*) Increase of beam increases initial stability, but has very little effect on range.

(*c*) Raising the centre of gravity decreases both initial stability and range.

(*d*) A ship which has negative initial stability will not necessarily capsize, but may become stable at some small angle of heel and may, thereafter, have a reasonable range of stability before she will capsize, provided that she has sufficient freeboard.

It must be remembered that the curves shown are for one particular case and are intended as a demonstration only. In practice, the average merchant ship often has a larger range of stability than that shown, but the conclusions that we have drawn will hold good in almost all cases.

Placing of Weights.—The naval architects who design a ship, make sure that she will be reasonably safe if she is properly loaded, as regards both her statical stability and her range of stability. They can, however, only fix the position of the centre of gravity for the ship when she is in her light condition. Its position during and after the loading of cargo will depend on the distribution of the weights, which is the duty of the ship's officers. It has already been seen that both the statical stability and the range of stability depend partly on the position of the centre of gravity, so those who load the ship must always remember that the final responsibility is on them.

It is not always possible to load ships exactly as we would wish, since we do not control the kind of cargo we receive, or the order in which it comes alongside. Thus, we sometimes have to "make the best of a bad job"; but even in the worst cases we can do quite a lot to control the stability of our ships by the judicious distribution of weights. If the seaman loads his ship so that she has a reasonably large metacentric height, he need not worry unduly about the range of stability, since the naval architects can be relied on to do their part of the work faithfully. In practice, the average merchant ship, when properly loaded and with a sufficient metacentric height, usually has a range of at least sixty to seventy degrees. Many still have a large righting lever even at ninety degrees of heel.

A "rule of thumb" method sometimes used at sea, is to place about one-third of the weight in the 'tween decks and two-thirds in the holds. This is a reasonably safe rule in most cases, but it must be remembered that all ships have their peculiarities and what is good for the average ship is not necessarily good for every one. The only truly reliable method is that of calculating the metacentric height.

Stiff and Tender Ships.—A stiff ship is one which has a large metacentric height (GM). A tender ship is one with a small metacentric height. These terms are relative: a ship does not suddenly become either stiff or tender at a given GM, but changes almost imperceptibly from one condition to another.

A good metacentric height for a fully-loaded merchant ship is usually between one half and one metre. A ship with a GM of less than this will normally be rather tender.

It is difficult to say just when a ship becomes stiff. A *GM* which would render one ship too stiff might be quite allowable in another: also, generally speaking, much larger *GM*s are considered reasonable in modern ships than would have been regarded as permissible some thirty years ago. It is probably fair to say, however, that a loaded ship with a *GM* of over one metre has a tendency to stiffness: whilst if her *GM* is much greater than this she will probably be too stiff. As we shall see later, a ship in the light condition normally has a large *GM*; often as much as from two to four metres.

Stiff Ships.—If a ship is too stiff, she will have an excessive righting moment and will tend to right herself violently when inclined. Her period of roll may be rather small and she will be liable to roll heavily and quickly in a seaway. This will cause her to be uncomfortable at sea and there is a risk that she may strain herself, or may cause her cargo to shift or to be damaged. Such a condition is not usually dangerous, but should be avoided whenever possible, for obvious reasons.

One is sometimes asked if it is advisable to pump out double bottom-tanks in a stiff ship. In port, this would be perfectly safe and good practice, subject to the ship being left with sufficient ballast for seaworthiness, since the metacentric height would thus be decreased. Whilst the tank is being pumped out, free surface effect would cause the centre of gravity to rise somewhat above its final position, but this should do no harm in the circumstances.

It would probably be safe to work tanks at sea in the same way, but it is not usually considered good practice to do so, unless absolutely necessary; because of the risk of structural damage to the tank due to free water washing about.

Tender Ships.—A tender ship will have a small righting moment and a comparatively long period of roll. She will have an easy motion in a seaway and may be quite safe, provided that her *GM* and freeboard are sufficient to give her an adequate range of stability.

This does not mean that it is good practice for a ship to be in a very tender state: on the contrary, such a condition should be avoided as much as undue stiffness. It is important to remember that the consumption of fuel and stores during a voyage usually causes the ship's centre of gravity to rise, so that she will probably arrive in port with a smaller *GM* than that with which she set out. If the ship is tender to begin with, this may cause her to become more so and she may even develop a negative *GM* before she finishes her voyage.

If a ship should become tender, the best cure is to work down weights and/or to fill double-bottom tanks, in order to lower the centre of gravity. Whilst tanks are being filled, free surface effect will cause *G* to rise slightly and the ship to become more tender, but this will disappear when the tank becomes nearly full. In modern ships the double-bottoms are usually well divided-up by the centre-line and side girders, which will reduce free surface even though they are not necessarily watertight. It is advisable to fill tanks which have a watertight division at the centre-line, in such cases.

Unstable Ships.—In British law, Masters and Officers are expected to make sure that their ships will be stable at all times. The fact that a ship was stable at the outset of a voyage is not accepted as extenuation, should she suffer loss or damage through becoming unstable during that voyage, unless the circumstances are extraordinary and unforeseeable. Apart from this, good seamanship demands that a ship should be so loaded as to ensure that the *GM* will not become negative at any time. Ships do sometimes become unstable, however, and if this happens, every effort must be made to rectify matters. The necessary steps are simple and obvious to those who understand stability, but unfortunately the wrong thing has sometimes been done occasionally with serious results.

Let us consider the causes of instability in practice:—

(*a*) Weights may be placed too high in the ship when cargo is being loaded and thus cause a negative *GM* at the outset of the voyage.

(*b*) Fuel and stores consumed at sea are often taken from a position low down in the ship. This may cause *G* to rise sufficiently, in a tender ship, to give her a negative *GM*.

(*c*) When a deck cargo is carried, it may soak up water during the voyage and thus increase its weight. This may also cause *G* to rise sufficiently to give the ship a negative *GM*.

(*d*) Water or oil may be taken from double-bottom tanks at sea. In this case, there will be a rise of *G* due to the removal of weight from the bottom and a further rise due to free surface effect in the slack tanks. This is particularly liable to happen in ships carrying oil fuel in the double-bottoms. The free surface effect is then unavoidable, but the other effect can be minimised by filling each tank with water ballast as it becomes empty.

It is obvious from the above that the ultimate cause of instability is a negative *GM* and that the only cure is to lower the ship's centre of gravity. This must be done very carefully, since it is always a difficult operation and it can actually be dangerous.

Weights such as bunkers, stores and cargo must, if possible, be trimmed down and/or double-bottom tanks filled, whilst in extreme cases it may become necessary to jettison cargo. The best general rule is to *add weight as low as possible or to remove it from the highest possible point.*

The most common and practical method of curing instability in a ship is to fill a double-bottom tank or tanks, but this may be dangerous if it is not done properly. Tanks which are divided at the centre-line should always be filled first, in order to minimise the effect of free surface. One tank should be filled at a time, commencing with the low side and when this is about two-thirds full, it will be safe to start running-up the high side. Free surface effect and the added weight on the low side will probably cause the ship to increase her list at first, but as the tank fills, she will gradually come upright. The high side of a tank should never be filled first, even though it may eventually achieve the desired result. There are two reasons against this: that *G* will not be lowered so quickly as by filling the low side first: that at some time the added weight on the high side will cause the ship to change her list, suddenly and violently, from one side to the other.

It is extremely dangerous and worse than useless to pump out double-bottom tanks in an attempt to correct a list. It might seem at first sight that if we pump out a tank on the low side of the ship, the removal of weight from that side would allow her to right herself. We must remember, however, that an unstable ship lists because her centre of gravity is too high and if we remove weight from the bottom of the ship, we shall only cause *G* to rise still higher. This rise of *G* will probably be aggravated by free surface effect. When sufficient weight has been removed from the low side of the tank, the ship will give a sudden "heave" and develop an even greater list to the other side; or she may even capsize.

A ship rarely becomes unstable when all the double-bottom tanks are full, but if this does occur, it is obvious that they should on no account be pumped out. If the ship is still dangerously unstable in such a case, after all possible cargo, fuel and stores have been shifted downwards, the only resort is to jettison cargo. When this is done, the cargo should first be taken from the high side of the ship and levelled off later. The reasons for this are the same as those for filling double-bottom tanks on the low side first.

Ships in Ballast.—When a ship has to make a voyage with no cargo on board, it is usually advisable to carry a certain amount of ballast. This makes the ship more seaworthy generally and immerses the propeller more deeply, thus increasing its efficiency and decreasing vibration. Modern ships use water ballast carried in tanks for this purpose.

A ship which is light usually has a large *GM* and is often excessively stiff. When water ballast is loaded into such ships it is important not to increase the *GM* further; and it is better, if possible, to reduce the *GM*.

It can be seen from the hydrostatic curves in the back of this book that, near the light draft, *M* falls quickly as draft and displacement increase. This is all to the good, since it tends to reduce the *GM* if we load water ballast. But if we load the ballast in double-bottom tanks alone it will cause *G* to fall considerably, so that the nett result is usually an increase in *GM*.

In order to avoid this, most ships have deep tanks, which can carry a large amount of water ballast higher up in the ship. Loading water into these will not lower the ship's centre of gravity appreciably. In this case, *M* will fall much more than will *G* and the nett result is usually a decrease in *GM*.

To illustrate this, consider a ship which, when light, has a draft of 3·20 m, displacement of 4986 t, *KG* of 6·50 m, *KM* of 10·14 m: and hence a *GM* of 3·64 m.

Now consider what will happen if we load 1500 t of water ballast: (*a*) in double bottom tanks, with their centres of gravity at 0·60 m above the keel: (*b*) in a deep tank, with its centre of gravity at 5·00 m above the keel. Suppose that, in each case, the new draft has become 4·00 m and the new *KM* is 9·02 m.

To find the new *GM*:—

		For (*a*)		For (*b*)	
	Weight	Height	Moment	Height	Moment
Ship	4,986	6·50	32,409	6·50	32,409
Ballast	1,500	0·60	900	5·00	7,500
Total	6,486		33,309		39,909
	New *KG* =	$\frac{33,309}{6,486}$	= 5·14 m	$\frac{39,909}{6,486}$	= 6·15 m
	KM =		9·02 m		9·02 m
	GM =		3·88 m		2·87 m

In this case, the water ballast, when loaded in the double bottom, will increase the *GM* by 24 cm: when loaded in the deep tank, it will decrease the *GM* by 77 cm.

The Effect of "Winging-Out" Weights. — "Winging-out" means placing weights well out from the centre-line towards the sides of a ship. Most seamen know that a ship so loaded is steadier in a seaway than one in which the heaviest weights are concentrated at the centre-line, all other things being equal.

A ship's period of roll depends largely on her moment of inertia. We have seen, in Chapter 3, that the greater the moment of inertia of a see-saw, the less quickly will it swing. Similarly with a ship; if her moment of inertia is increased, her period of roll will also become greater. If the weights in the ship are winged well out, they will cause her to have a greater radius of gyration than she would have if they were near the centre-line. This will increase her moment of inertia and period of roll, so that she will be steadier in a seaway.

It must be remembered, to avoid confusion, that we are here considering the moment of inertia of the ship herself; not that of the waterplane, as we did when we were finding *BM*.

Deck Cargoes.—Ships carrying heavy deck cargoes are always liable to become unstable, since the additional weight is placed high in the ship. If the cargo is of a type which is likely to soak up water during the voyage, the consequent increase of weight on deck may cause *G* to rise sufficiently to make the vessel unstable. When such a cargo is being loaded, therefore, a sufficient margin of safety must be allowed for this eventuality.

Timber Deck Cargoes.—The remarks made in the last paragraph also apply to the particular case of a timber deck cargo. When such a cargo is properly secured, however, it becomes in effect an addition to the ship's hull and thus increases the freeboard. We have seen that an increase in freeboard will increase the range of stability so that a ship carrying a timber deck cargo may be perfectly safe, even though she is tender. In Fig. 61, curves *E* and *F* shew that such a ship may even have a small list on account of the extra weight on deck, and yet have quite a large range of stability. This would be bad seamanship, but not necessarily dangerous as far as stability is concerned.

The advantage of increased range of stability can obviously only be gained if the deck cargo is efficiently secured so as to form a solid block with the ship's hull. It is worth noting that the regulations with regard to deck cargoes of timber carried on ordinary ships lay down that such cargo must be compactly lashed, stowed and secured and that it must not render the vessel unstable during the voyage.

The Load Line Rules require that an allowance equal to 15 per cent of the weight of a timber deck cargo shall be made for water soaked up by the timber.

Ships marked with lumber load-lines are allowed to load more deeply when carrying a timber deck cargo, than at other times. Since the additional weight will normally be on the deck in such cases, it is important that the stability of these ships should be even more carefully considered. Three points from the regulations with regard to this are worth noting particularly:—

(*a*) The double-bottoms must have adequate longitudinal sub-divisions. This is obviously a precaution against undue free surface effect when the tanks are filled, to prevent the ship from becoming unduly tender.

(*b*) The timber must be stowed solid to a certain minimum height. This ensures sufficient freeboard to give an adequate range of stability, if the ship becomes very tender. It also means that if she were to lose the deck cargo, she would rise approximately to her ordinary load-lines.

(*c*) The lashings have to conform to very stringent rules, which ensure that the deck cargo forms a solid mass with the ship.

Free Liquid in Tanks.—The importance of longitudinal subdivisions in tanks has been referred to several times. A study of Chapter 9 will shew that the smaller the area of free surface in a tank, the less will be the rise of the ship's centre of gravity due to such surface; also that a decrease in its breadth will have a much greater effect than a decrease in length. Hence, the best way of minimising the effect is to use a tank which has as many longitudinal subdivisions as possible. Washplates are quite as effective as watertight subdivisions for this purpose, provided that they extend to below the surface of the liquid. The modern cellular double-bottom tank has, at least, a watertight centre girder and two side girders, which will act as washplates, so that the free surface is divided into at least four parts. Slack double-bottom tanks should always be avoided if possible but they will not usually be dangerous, unless the ship is very tender.

The amount of liquid in a tank will not appreciably affect the position of the virtual centre of gravity due to free surface, unless it changes the shape of that surface. The weight of the liquid does, however, affect the final position of the ship's centre of gravity for two reasons. In the first place, it will have an influence on the original position of *G*. Secondly, it will change the volume of displacement of the ship and will thus cause a slight change in the rise of *G* due to free surface. In theory, one centimetre of water in a double-bottom tank would cause the centre of gravity of the ship to rise much higher than, say, one metre of water: the free surface effect would be the same in each case, but in the second case, the original centre of gravity would be lower, on account of the extra weight in the bottom of the ship. This would hold good in practice as long as the ship were perfectly upright, but as soon as she heeled slightly, the water would run down into one corner. If the tank were nearly empty, or nearly full, this would cause a considerable decrease in the free surface.

There is always a large free surface effect when deep tanks are being filled. This is not normally dangerous, since, in the average ship, such tanks are only filled when she is light and, therefore, comparatively stiff. Some modern ships carry liquid cargoes and/or bunkers in deep tanks and peak tanks, however, and may only have a small metacentric height when such tanks are filled. In this case, free surface effect becomes important and must be considered carefully.

Free liquid in tanks, as distinct from pure free surface effect, is not usually considered, because it has peculiar and apparently unpredictable effects on the rolling of ships. There is no doubt, however, that the period of surge of the liquid is sometimes the same as the ship's period of roll and when this happens, it increases the rolling.

Apart from any question of stability, it must be remembered that slack tanks are always bad from a structural point of view. Free liquid exerts a considerable lifting effect on the tank tops and may cause considerable damage to them.

It can be seen from the above that free liquid in tanks is always objectionable and should be avoided whenever possible, even when the free surface effect is not dangerous. When it does occur, one should keep a sense of proportion and neither underestimate nor overrate its possibilities.

Free Surface Effect in Oil Tankers.—This effect presents a special problem in the case of oil tankers, since, when tanks are "full", a certain amount of space (or "ullage") must be left between the surface of the oil and the tank-top to allow for expansion of the cargo due to changes of temperature. The usual methods of minimising free surface effect, in this case, are shown diagrammatically in Figs. 62 (*a*) and 62 (*b*).

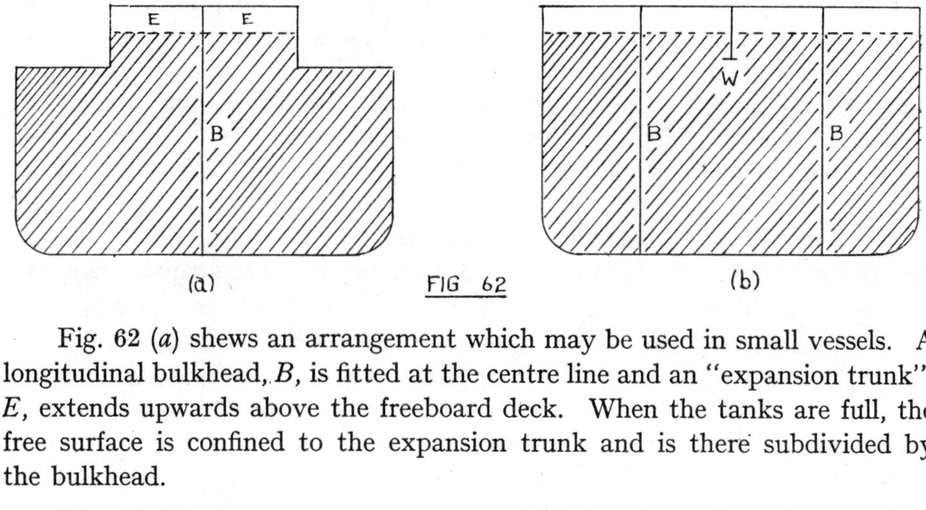

FIG 62

Fig. 62 (*a*) shews an arrangement which may be used in small vessels. A longitudinal bulkhead, *B*, is fitted at the centre line and an "expansion trunk", *E*, extends upwards above the freeboard deck. When the tanks are full, the free surface is confined to the expansion trunk and is there subdivided by the bulkhead.

Fig. 62 (*b*) shows an arrangement which is often adopted in modern tankers. There is no expansion trunk, but two longitudinal bulkheads *B* are fitted, one at each quarter line and also a washplate *W*, or a non-watertight bulkhead, at the centre-line. When the tank is partly full, the free surface is divided into three nearly equal parts; when it is nearly full, so that the washplate extends to below the surface of the oil, that surface is divided into four parts.

CHAPTER 11

DYNAMICAL STABILITY

Definition.—Dynamical stability is the amount of work done in inclining a ship to a given angle of heel.

Work.—Suppose that we wish to push a weight across the deck of a ship. The weight will resist our efforts to move it on account of inertia, friction with the deck, etc., and we shall have to exert force in order to start it moving. If we then stop pushing, the friction between the deck and the weight will soon cause the latter to stop moving, so we must continue to push until it is in the desired position. The greater the weight, the harder we must push and the greater the distance, the longer we must push. In other words, we must do work and the amount of work done depends on the distance we have to move the weight and the amount of force we have to exert in order to move it. Thus, work done is equal to the force exerted, multiplied by the distance over which it is exerted.

Dynamical Stability.—Consider a ship which is being heeled by some external force. As soon as she heels to a small angle, her moment of statical stability will try to force her back to the upright. In order to heel her further, sufficient force must be exerted to overcome this statical stability and must continue to be exerted for as long as the ship continues to heel. We can liken this case to that of the weight mentioned in the last paragraph and say that the work done to heel the ship to any given angle is equal to all the force exerted, over all the distance through which the ship has heeled. This is obviously only another way of expressing the definition of dynamical stability, which is given above.

Dynamical Stability from a Curve of Statical Stability.—Fig. 63 shows a curve of statical stability, in which the moment of statical stability ($W \times GZ$) is plotted against the angle of heel. The statical stability at any angle is found by the perpendicular distance from the base line to the curve at that angle. For instance, the moment of statical stability at 30° of heel is found by drawing the perpendicular line, AB, and then the horizontal one, BD. The required moment is then CD—in this case about 13,000 tonne-metres.

FIG. 63

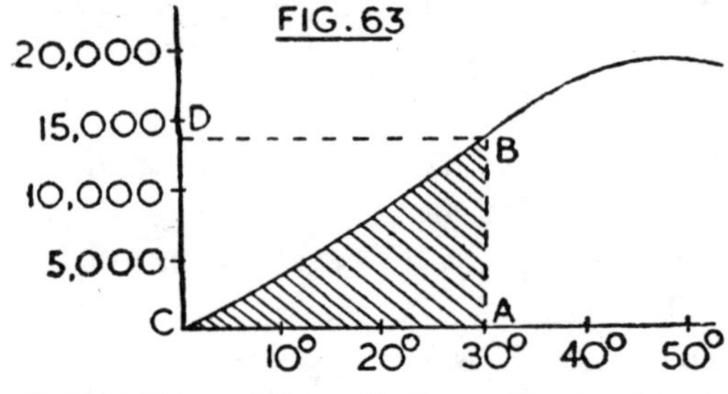

We have just said that dynamical stability is equal to all the force exerted over all the distance through which the ship has heeled. This can be taken to mean the sum of all the moments of statical stability, for every minute angle of heel, up to the given angle. If we consider Fig. 61 again, we shall see that the sum of all the moments of statical stability up to 30° of heel, will be equal to the shaded area ABC. Similarly for other angles. This means that the *dynamical stability at any angle is equal to the area under the curve of statical stability up to that angle.* For this purpose, the vertical distances to the curve are always measured in terms of statical stability and the length along the base line in terms of circular measure (or "radians"). A radian is equal to 57·3°, so that the length along the base line becomes:—

$$\frac{\text{Angle of heel, in degrees}}{57{\cdot}3}$$

Thus, given a scale or curve of righting levers or moments, we can use Simpson's Rules to find the Dynamical Stability.

Strictly speaking, we should always use moments of statical stability ($W \times GZ$) to obtain dynamical stability, which should be measured in "tonne metre radians". For purposes of comparison, however, we often use righting levers (GZ), to give a quantity called "metre radians". The latter can be converted to actual dynamical stability by multiplying it by the ship's displacement (W), as illustrated in the following example.

Example.—Calculate the dynamical stability at 40° heel, for a ship which displaces 6000 tonnes and which has the following righting levers:—

Heel	10°	20°	30°	40°
GZ	0·18	0·41	0·67	0·85

$$\text{Common interval} = 10° = \frac{10}{57{\cdot}3}\text{ radians}$$

Ordinate	Multiplier	Product
0·00	1	0·00
0·18	4	0·72
0·41	2	0·82
0·67	4	2·68
0·85	1	0·85
		5·07

$$\text{Area} = \frac{h}{3} \times 5{\cdot}07$$

$$= \frac{10}{57{\cdot}3} \times \frac{1}{3} \times 5{\cdot}07$$

$$= 0{\cdot}2949 \text{ metre-radians}$$

$$\text{Dynamical stability} = 0{\cdot}2949 \times \text{displacement}$$

$$= 0{\cdot}2949 \times 6000 = 1770 \text{ tonne-metre radians}$$

Calculation of Dynamical Stability.—Dynamical stability can, if desired, be found by direct calculation.

When a ship heels, the centres of gravity and buoyancy separate vertically G moves slightly upwards with the ship, whilst B moves outward and downwards, because its shift is parallel to the shift of the centres of gravity of the immersed and emerged wedges. The upward shift of G is resisted by the weight of the ship, which acts vertically downwards through it. The downward shift of B is resisted by the upward thrust of the force of buoyancy, which is also equal to the weight of the ship. Thus, when a ship heels, B and G are forced apart against a resistance equal to the ship's weight, or displacement. The work done, or dynamical stability, is thus equal to the displacement, multiplied by the vertical separation of the centres of gravity and buoyancy.

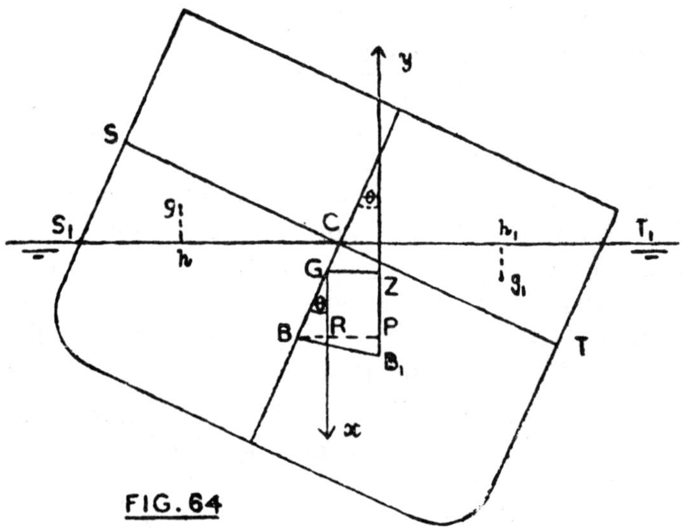

FIG. 64

Consider Fig. 64. When the ship was upright, the vertical distance between the centres of buoyancy and gravity was BG. With the ship heeled to the angle θ, B has moved out to B_1, and the vertical distance has become B_1Z. So the vertical separation of B and G, due to the heel of the ship is:—

$$(B_1Z - BG).$$

If W is the ship's displacement, this means that:—

$$\text{Dynamical Stability at the angle } \theta = W\,(B_1Z - BG).$$

Moseley's Formula.—This is a development of the above formula. Consider Fig. 64 again:—

Let g and g_1 be the centres of gravity of the emerged and immersed wedges. Drop the perpendiculars gh and g_1h_1 on to the new waterline, S_1T_1.

Let W = the ship's displacement.
V = her volume of displacement.
v = the volume of the immersed wedge or of the emerged wedge (SCS_1 or TCT_1).
θ = the angle of heel.

It can be shewn that:—

$$\text{Dynamical stability} = W\left\{\frac{v\,(gh + g_1h_1)}{V} - BG \times \text{versine}\ \theta\right\}$$

Notes.—Dynamical stability is important in ship stability for two reasons. Since it is the measure of the work that must be done to heel a ship:—

(*a*) It is a big factor in deciding how a ship will roll; in this case the waves are doing the work.

(*b*) It determines the ability of a sailing ship to stand-up under sail; the pressure of the wind on the sails supplying the work in this case. This is not of much interest to the average merchant seaman today, but is important to yachtsmen and others who have to deal with sailing craft.

It must always be remembered that anything which reduces statical stability will also reduce dynamical stability. This is yet another count against our old enemy, the free surface of liquids.

CHAPTER 12

LONGITUDINAL STABILITY

Recapitulation.—It is now time to consider the effects of what may be called the longitudinal stability of a ship. Before we begin, it will be as well to refresh our memories on certain matters which were discussed in previous chapters.

The centre of gravity (G) and centre of buoyancy (B) will not necessarily be amidships, but may be forward or aft of this.

The centre of flotation (F) is the centre of gravity of the ship's waterplane and is the point about which the ship heels and trims. In box shapes, it is always amidships: in ship shapes it may be a little abaft or forward of the centres of buoyancy and gravity. Its longitudinal position changes with change of draft and sometimes with change of trim also. It is sometimes called the "tipping centre".

The longitudinal metacentre (M_L) is a different point from the transverse metacentre, although it is found in a similar way and obeys similar laws.

The longitudinal metacentric height (GM_L) is always very large, often well over one hundred metres. If the position of G is not known, BM_L can often be used instead of GM_L, the error thus caused being negligible in practice.

Draft is the depth of the bottom of the keel below the waterline. It is marked and measured forward and aft at the stem and the rudder post, respectively.

Mean draft is the mean of the drafts fore and aft: that is, the draft amidships in the way of the ship's loadlines, midway between the stem and the rudder post.

$$\text{Mean draft} = \frac{\text{Draft forward} + \text{Draft aft}}{2}$$

Tonnes per centimetre immersion (T.P.C.) is the weight required to cause the ship to rise or to sink bodily by 1 centimetre.

Longitudinal Metacentric Height—"GM_L".—The calculations to find this are very similar to those used for transverse metacentric height. BM_L is first calculated, whilst KB and KG will have already been found for the transverse calculations, so that GM_L can be found by addition and subtraction.

For instance, suppose that a ship's BM_L is found by calculation to be 130·4 metres, that KB is 3·3 metres and KG is 5·8 metres. Then:—

$$\begin{aligned} BM_L &= 130{\cdot}4 \text{ metres} \\ KB &= \underline{3{\cdot}3} \quad ,, \\ KM_L &= 133{\cdot}7 \quad ,, \\ KG &= \underline{5{\cdot}8} \quad ,, \\ GM_L &= \underline{\underline{127{\cdot}9}} \quad ,, \end{aligned}$$

The Calculation of "BM_L" for all Shapes.—Figure 65 shows a ship which has tipped longitudinally through a small angle, θ. B has moved out to B_1 and the longitudinal metacentre, M_L, has appeared.

Compare this figure with Fig. 54, Chapter 8, which shows the transverse metacentre appearing as the ship heels. In each case there is an emerged and an immersed wedge; a shift of B to B_1, parallel to the shift of the centres of

FIG. 65

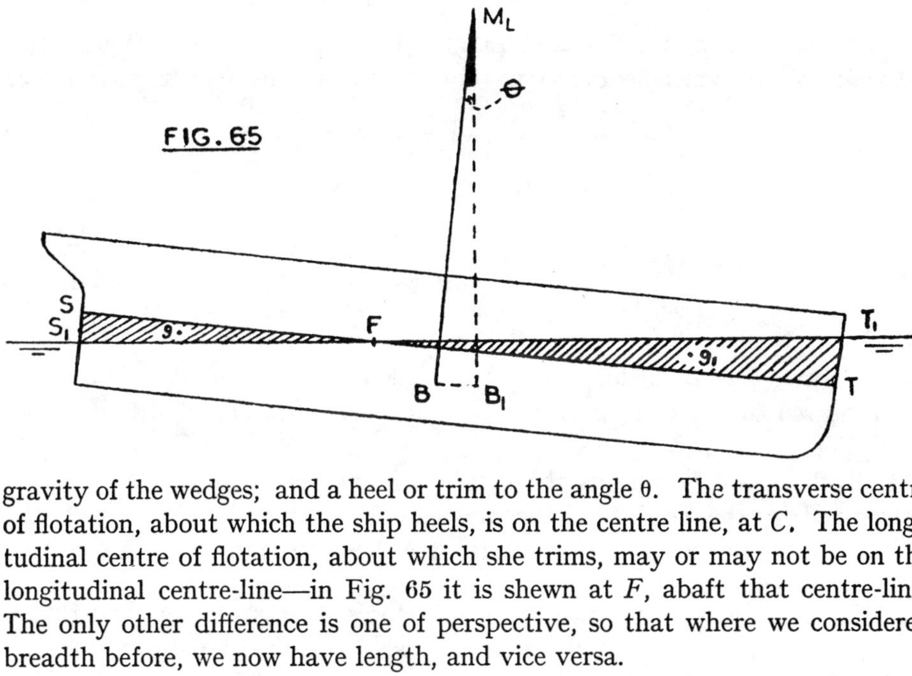

gravity of the wedges; and a heel or trim to the angle θ. The transverse centre of flotation, about which the ship heels, is on the centre line, at C. The longitudinal centre of flotation, about which she trims, may or may not be on the longitudinal centre-line—in Fig. 65 it is shewn at F, abaft that centre-line The only other difference is one of perspective, so that where we considered breadth before, we now have length, and vice versa.

It can be seen from the above that we could show by means of a proof similar to that for transverse BM, that:—

Where I_L is the moment of inertia of the waterplane, longitudinally, about the centre of flotation, and V the ship's volume of displacement; then for all shapes, including ship shapes,

$$BM_L = \frac{I_L}{V}$$

The Calculation of "BM_L" for Box Shapes.—In the case of box shapes, the centre of flotation is on the longitudinal centre-line of the ship. If we consider the moment of inertia of a rectangle about its centre-line, it follows that if l is the ship's length, b her breadth and D the draft at which she floats:—

$$I_L = \frac{bl^3}{12}$$

but $$BM_L = \frac{I_L}{V}$$

so $$BM_L = \frac{bl^3}{12V}$$

but $V = l \times b \times D$, so the above becomes:—

$$BM_L = \frac{bl^3}{12l \times b \times D}$$

$$BM_L = \frac{l^2}{12D} \quad \text{(for box shapes only)}$$

Example.—Find the BM_L and GM_L of a box-shaped vessel, 120 metres long, 15 metres beam, which floats at a draft of 4·00 metres, the KG being 5·00 metres

$$BM_L = \frac{l^2}{12d} = \frac{120^2}{12 \times 4} = 300 \text{ metres}$$

$KB = \frac{1}{2}$ draft	= 2	,,
KM_L	= 302	,,
KG	= 5	,,
GM_L	= 297	,,

Trim.—This is the longitudinal equivalent of heel, but whereas the latter is measured in angle, trim is measured by the difference of the drafts fore and aft.

If the draft forward is the greater of the two, the ship is said to be "trimmed by the head". If the draft aft is the greater, she is said to be "trimmed by the stern". If the drafts are the same, fore and aft, she is said to be "on an even keel".

In many ways, the calculation of a ship's trim is simpler than that for heel, but there is one complication that we do not meet with in transverse calculations. A ship heels and trims about her centre of flotation and when she is upright, the transverse positions of the centres of gravity, buoyancy and flotation are all vertically over one another, on the centre-line. Longitudinally, none of them will necessarily be amidships; and further, although B and G must be in the same vertical line, the centre of flotation is very rarely directly over them. We shall shortly consider the effect of this on trim and sinkage due to added weights.

Change of Draft due to a Change of Trim.:—When a ship changes her trim, she can be considered to increase her draft at one end and to decrease it at the other. The sum of the changes at both ends is the change of trim assuming that there is no increase of draft due to added weights.

The change of draft due to change of trim will depend on the position of the centre of flotation. When this is at the longitudinal centre-line, the ship will increase her draft at one end by exactly half the change of trim and will decrease it by a like amount at the other end. The mean draft will not change.

When the centre of flotation is not amidships, the draft will change more at one end than at the other, because the ship will be tipping about a point which is not midway between the ends. In this case, the change of draft can be found by a simple proportion.

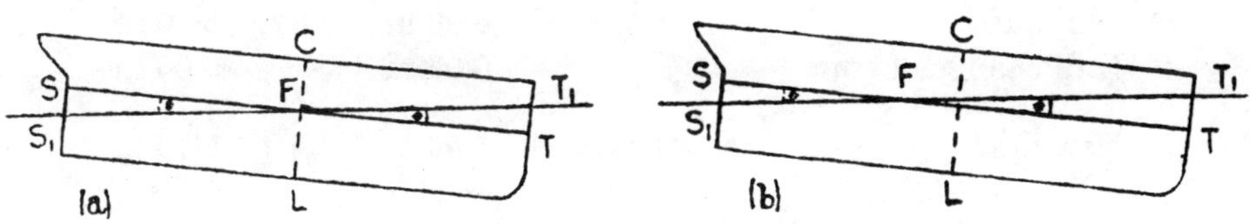

FIG. 66

Figures 66 (*a*) and (*b*) show what happens in the above cases. *F* represents he centre of flotation and *CL* the centre-line of the ship, whilst *ST* and S_1T_1 are the old and new waterlines, respectively. In each case the change of trim is $SS_1 + TT_1$.

In Fig. 66 (*a*), *F* is amidships It can be seen that, for all practical purposes, *SF* equals *TF*, S_1F equals T_1F, whilst the angle θ is the same on either side of *F*. The triangles SS_1F and TT_1F are, therefore, equal in all respects, so:—

$$SS_1 = TT_1$$

$$\text{But } SS_1 + TT_1 = \text{the change of trim}$$

$$\text{So, } SS_1 = TT_1 = \text{half the change of trim.}$$

This means that when the centre of flotation is amidships the change of draft at either end is equal to half the change of trim.

Fig. 66 (*b*) shows what happens when *F* is not amidships. In the triangles SS_1F and TT_1F the angles θ are equal and the angles at *S* and *T* may be considered to be right angles, so that the triangles are similar but not equal. One of the properties of similar triangles is that the sides are proportional to each other, so:—

$$SS_1 : TT_1 :: SF : TF$$

Now, $SS_1 + TT_1$ = the change of trim (call this t).
and $SF + TF$ = the length of the ship on the waterline (call this l)

So, $$SS_1 = \frac{SF}{l} \times t$$

and $$TT_1 = \frac{TF}{l} \times t$$

Example 1.—A box-shaped vessel has its centre of flotation amidships and floats at drafts of 5·00 metres forward and 5·16 metres aft. A weight is shifted forward and causes a change of trim of 12 centimetres. Find the new drafts.

Since the centre of flotation is on the centre-line, the draft at each end will change by half the change of trim. The draft forward increases and that aft decreases, so:—

Old drafts - - -	Forward	5·00 m	Aft	5·16 m
Half change of trim -		+0·06 m		−0·06 m
New drafts - - -	Forward	5·06 m	Aft	5·10 m

Example 2.—A ship is 120 metres long and her centre of flotation is 3 metres abaft amidships. She floats at drafts of 7·45 metres forward and 7·33 metres aft. Find the new drafts if a weight is shifted aft so as to change the trim by 56 centimetres.

The centre of flotation is 63 metres from forward and 57 metres from aft.

$$\text{Change of draft forward} = \frac{63}{120} \times 56 = 29 \text{ cm}$$

$$\text{Change of draft aft} = \frac{57}{120} \times 56 = 27 \text{ cm}$$

Original drafts - -	F.	7·45 m	A.	7·33 m	M.	7·39 m
Change due to trim -		−0·29 m		+0·27 m		
New drafts - - -	F.	7·16 m	A.	7·60 m	M.	7·38 m

Note that in the above case, although no weight has been added, the mean draft has decreased by one centimetre. The cause of this is discussed later.

Change of Mean Draft due to Change of Trim.—When the centre of flotation is not amidships, the mean draft will change slightly with change of trim. This will happen, even if a weight has merely been shifted, not added or removed, so that the ship's displacement has not changed. Sometimes this change of mean draft is so small as to pass unnoticed, but at other times it may amount to a centimetre or more, as we saw in the last example.

At first sight this might appear impossible, since the ship's displacement curves give only one mean draft for a given displacement. The displacements given in the curves are, however, only calculated for one condition of trim: the "designed trim", which is usually for waterlines parallel to the keel. When the ship changes her trim, she tips about the centre of flotation: if this is not amidships, where the mean draft is measured, the latter must change as the ship trims.

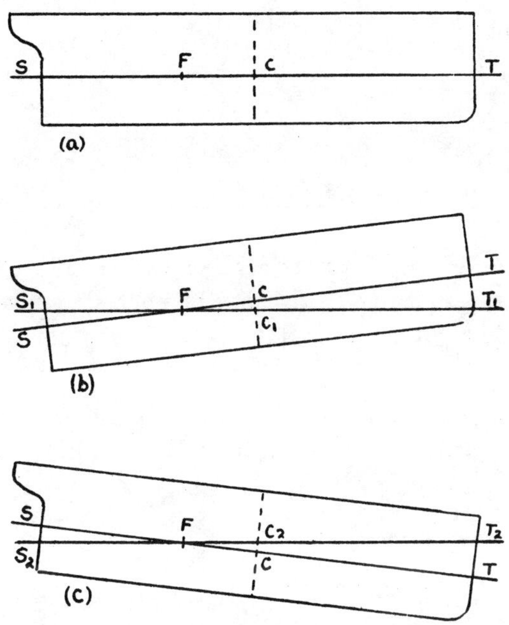

FIG. 67

Fig. 67 (a) shows the ship floating at her designed trim, at the waterline ST. C is the point amidships at which the mean draft is taken: F the centre of flotation, abaft C.

In Fig. 67 (b), a weight has been shifted to cause the ship to trim by the stern. She has tipped about F to the new waterline S_1T_1, which cuts the centre-line at C_1, the new mean draft. The mean draft has decreased from C to C_1, although the displacement has not changed.

Fig. 67 (c) shows the same ship trimmed by the head. The new waterline is now S_2T_2 and the mean draft has increased to C_2, for the same displacement.

We can convert mean draft to draft at the centre of flotation, if necessary, by a simple calculation:—

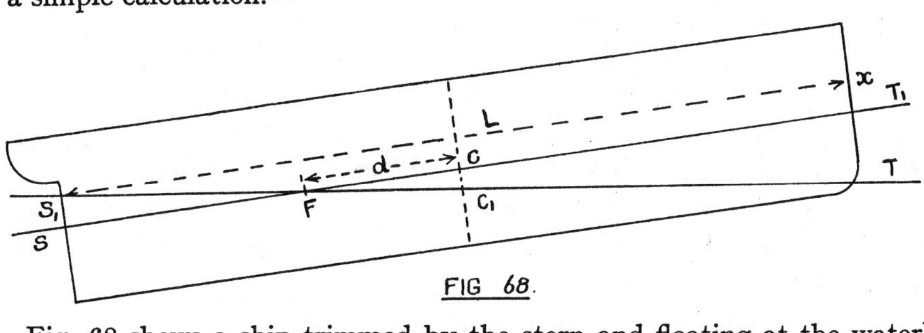

FIG 68.

Fig. 68 shews a ship trimmed by the stern and floating at the waterline S_1T. ST_1 is the equivalent even-keel waterline. The even-keel drafts are at C

and F, whilst the mean draft is at C_1. The ship's trim is equal to $SS_1 + TT_1$. The line S_1x is drawn parallel to ST_1, so that S_1x is equal to the ship's length (L.). Let the distance of F from amidships (FC) be d metres.

$Tx = SS_1 + TT_1 =$ ship's trim.

The triangles S_1xT and FCC_1 are similar triangles.

$$\frac{CC_1}{Tx} = \frac{FC}{S_1x}$$

$$\frac{CC_1}{\text{Trim}} = \frac{d}{L}$$

So, $$CC_1 = \frac{d \times \text{Trim}}{L}$$

If the designed trim is not for even-keel drafts, we can modify this to:—

$$CC_1 = d \times \frac{\text{Amount out of designed Trim}}{L}$$

If F is on the same side of amidships as the end which is "down", the draft at F will be greater than the mean draft. If F is on the opposite side of amidships to the end which is "down", the draft at F will be less than the mean draft.

Example.—A ship is 120 m long and floats at drafts of 6·76 m forward and 7·48 m aft. What is the equivalent even-keel draft, if F is 3·2 m abaft amidships?

$$\text{Mean draft} = \frac{6{\cdot}76 + 7{\cdot}48}{2} = 7{\cdot}12\text{m}$$

$$\text{Trim} = 7{\cdot}48 - 6{\cdot}76 = 72\text{cm (by the stern)}$$

$$\text{Difference of drafts } (CC_1) = \frac{3{\cdot}2 \times 72}{120} = 1{\cdot}9\text{cm}$$

$$\text{Even-keel draft} = \text{mean draft} + CC_1$$
$$= 7{\cdot}12 + 0{\cdot}02 = 7{\cdot}14\text{m}$$

Displacement out of the Designed Trim.—We can see from the above that a ship's displacement may vary slightly for a given mean draft, according to her trim. The displacement scale or curves will only give the correct displacement for the designed trim: that is for the draft at F. This means that if we use the mean draft to obtain displacement from the scale or curve, there may be a small error in the displacement so found. In most cases, this would not matter; but if it is necessary to obtain the displacement with greater accuracy, we can do so in one of two ways:—

(*a*) Convert the mean draft to draft at the centre of flotation, as above. Then take out the displacement for the latter from the scale or curves.

(*b*) Take out the displacement for the mean draft from the scale or curve. Then apply a correction, called a "layer correction", to that displacement. The method of finding this correction is as follows:—

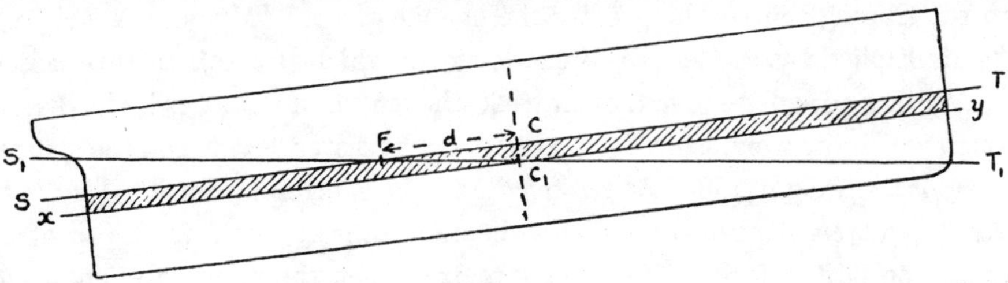

FIG. 69

In Fig. 69, *ST* represents the designed waterline for the draft at *F* or *C*. *xy* is the designed waterline for the mean draft, at C_1. The shaded area between the two waterlines is the layer between them.

$$\text{Weight of such a layer} = \text{T.P.C.} \times \text{its depth in cms}$$

$$\text{So, layer correction} = \text{T.P.C.} \times CC_1 \text{ in cms}$$

$$\text{But } CC_1 = \frac{d \times \text{Amount out of Designed Trim}}{L}$$

$$\text{So, layer correction} = \frac{\text{T.P.C.} \times d \times \text{Amount out of Designed Trim}}{L}$$

This correction must be added to the displacement taken for the mean draft, if the end nearer to the centre of flotation is "down": or subtracted if the other end is "down".

Example.—A ship is 130 metres long, her T.P.C. is 13·0 tonnes and her centre of flotation is 2·0 metres abaft amidships. She floats at drafts of 6·82 metres forward and 7·18 metres aft. Her designed trim is an even keel. If the displacement for a mean draft of 7·00 metres is given in the scales as 8016 tonnes, find the true displacement.

The mean draft is 7·00 metres.

The ship is trimmed 36 cm by the stern. Since the designed trim is an even keel, this means that she is 36 cm out of designed trim.

$$\text{Layer correction} = \frac{\text{T.P.C.} \times d \times \text{Amount out of designed trim}}{L}$$

$$= \frac{13 \times 2 \times 36}{130} = 7 \text{ tonnes}$$

$$\text{Displacement from scale} = 8016 \text{ tonnes}$$

$$\text{True displacement} = 8023 \text{ tonnes}$$

Moment to Change Trim by 1 Centimetre—"M.C.T.1C."—If we add, remove, or shift a weight forward or aft, the ship will change her trim. The greater the weight, or the greater the distance it is shifted or placed away from the tipping centre, the greater the change of trim. Moment is weight, or force, multiplied by the length of the lever on which it acts. In this case the weight is that added, removed, or shifted; the length of the lever is the distance through which the weight has been moved, or the distance between the centre of gravity of the weight added or removed and the centre of flotation. If we add or remove a weight of w tonnes at a distance of x metres from the centre of flotation, or shift it for a distance of x metres, then the moment changing the trim is wx tonne-metres. *The moment to change trim by one centimetre is the moment which will change a ship's trim by exactly one centimetre.* Its amount will vary in the same ship with change of draft.

Fig. 70 represents a ship in which a weight, w, has been shifted forward to w_1, through a distance of d metres. B and G have moved to B_1 and G_1 respectively, and the longitudinal metacentre, M_L, has appeared. ST is the original waterplane and S_1T_1 the new waterplane, the ship trimming about the centre of flotation, F.

Let θ be the angle between the original and new waterplanes, W the ship's displacement and l her length.

FIG. 70

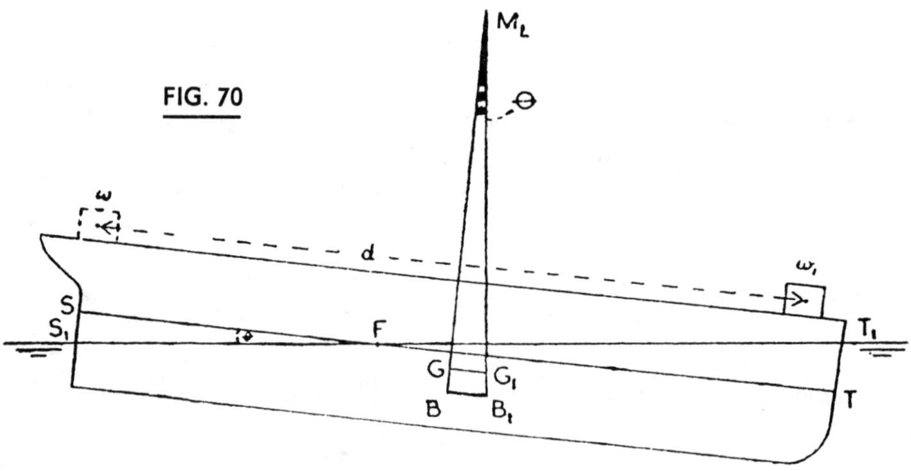

Let us do a little simple trigonometry:—In the triangle GG_1M_L, the angle G is equal to 90° and the angle M_L to θ. So:—

$$GG_1 = GM_L \times \tan\theta \qquad (1)$$

And yet more trigonometry:—In the triangle FTT_1, the angle T is equal to 90° and the angle F to θ. In the triangle FSS_1, the angle S is equal to 90° and the angle F to θ. So:—

$$TT_1 = FT \times \tan\theta$$
$$SS_1 = FS \times \tan\theta$$

Adding these, we get:—

$$TT_1 + SS_1 = FT \times \tan\theta + FS \times \tan\theta$$
$$= (FT + FS)\tan\theta$$

But $(TT_1 + SS_1)$ equals the change of trim, whilst $(FT + FS)$ is equal to the ship's length, so:—

$$\text{Change of trim} = l \times \tan\theta$$

If the trim is to change one centimetre—i.e., one-hundredth of a metre:—

$$l \times \tan\theta = \frac{1}{100}$$

$$\tan\theta = \frac{1}{100l} \qquad \qquad (2)$$

The change of trim is due entirely to the shifting of the weight, so we can say that:—

$$\text{Moment changing trim} = w \times d$$
$$\text{But,} \quad w \times d = W \times GG_1$$
$$\text{So, Moment changing trim} = W \times GG_1$$

Substituting for GG_1 from formula (1), we get:—

$$\text{Moment changing trim} = W \times GM_L \times \tan\theta$$

Substituting for $\tan\theta$ from formula (2), this gives:—

$$\text{Moment to change trim 1 cm} = W \times GM_L \times \frac{1}{100l}$$

$$\textit{i.e.}\ \text{M.C.T., 1c.} = \frac{W \times GM_L}{100l}$$

If GM_L is not known, we can substitute BM_L for it in the above formula without causing any appreciable error.

Example.—A ship is 120 metres long, has a displacement of 3618 tonnes and a GM_L of 180 metres. Find her M.C.T.1C.

$$\text{M.C.T.1C.} = \frac{W \times GM_L}{100l} = \frac{3618 \times 180}{100 \times 120} = 54{\cdot}27 \text{ tonne-metres}$$

The Effect of Shifting a Weight.—We have already seen that for a ship to be in transverse equilibrium, her centre of gravity must be vertically over her centre of buoyancy. This applies equally in the longitudinal direction.

If a weight is shifted forward or aft, the centre of gravity will move parallel to it and will no longer be vertically over the centre of buoyancy, unless the ship changes her trim.

This can be seen from Fig. 71 (*a*). The weight, w, has here been moved forward to w_1 and G has moved out to G_1. The force of gravity now acts vertically downwards through G_1, in the direction xG_1, whilst the force of buoyancy acts vertically upward through B, in the direction yB. The effect of this is to force the ship's head downwards, so that she starts to change her trim about the centre of flotation, F.

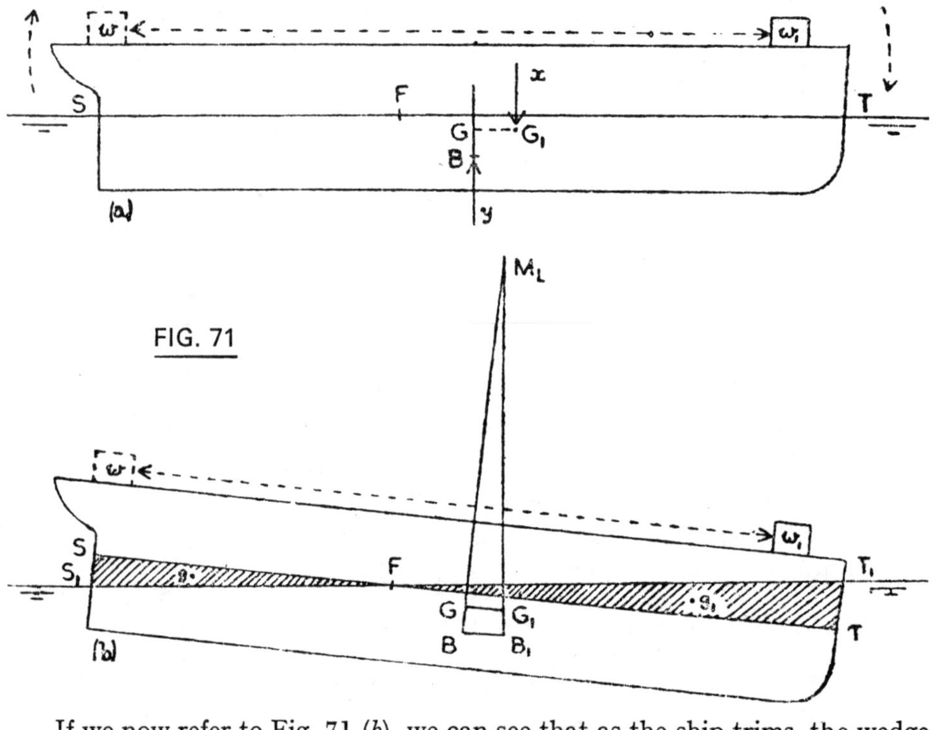

FIG. 71

If we now refer to Fig. 71 (*b*), we can see that as the ship trims, the wedge SFS_1 will emerge from the water and the wedge TFT_1 will become immersed. B will move out parallel to the shift of the centres of gravity of the wedges (g and g_1), until it reaches B_1, vertically under G_1. The ship is then in equilibrium and floats at the new waterline S_1T_1. Note how the longitudinal metacentre, M_L has appeared.

The moment changing the trim is the weight, in tonnes, multiplied by the distance, in metres, through which it has been shifted. Since M.C.T.1C. is the moment to change the trim by one centimetre, we can say that:—

$$\text{Change of Trim} = \frac{\text{Moment changing trim}}{\text{M.C.T.1C.}}$$

So, where w is the weight, in tons, and d the distance through which it is shifted forward or aft,

$$\text{Moment changing trim} = w \times d$$

$$\text{Change of trim, in centimetres} = \frac{w \times d}{\text{M.C.T.1C.}}$$

The change of draft due to this can be found from the change of trim, as shown in the last section. If the weight is shifted aft, the after draft will increase and the forward draft decrease; if the weight is shifted forward, the reverse will happen. If the position of the centre of flotation is not known, we usually assume it to be amidships since the error thus caused, if any, is generally small.

Example.—Find the change of trim and the new drafts in a ship of 125 metres length, when a weight of 100 tonnes is shifted aft for a distance of 60 metres. The original drafts were 4·80 metres forward and 4·76 metres aft. The centre of flotation is 2·5 metres abaft amidships and the M.C.T.1C. is 120 tonne-metres.

$$\text{Change of trim} = \frac{\text{Moment changing trim}}{\text{M.C.T.1C.}} = \frac{100 \times 60}{120}$$

$$= 50 \text{ cm}$$

F. is 65 m from forward and 60 m from aft.

$$\text{Change of draft forward} = \frac{65}{125} \times 50 = 26 \text{ cm}$$

$$\text{Change of draft aft} = \frac{60}{125} \times 50 = 24 \text{ cm}$$

Original drafts - -	F.	4·80 m	A.	4·76 m	M.	4·78 m
Change - - -		−0·26 m		+0·24 m		
New drafts - - -	F.	4·54 m	A.	5·00 m	M.	4·77 m

The Effect of Adding a Weight at the Centre of Flotation.—Suppose that, as in Fig. 72 (*a*), a ship floats at a waterline, ST, and that a weight, w is then added directly over the centre of flotation, F. Let B and G be vertically under F. When the weight is added, G will move vertically upwards, towards the centre of gravity of the weight. The ship will sink to the new waterline S_1T_1, so that the weight of the added layer, SS_1T_1T, is equal to that of the added weight. The new centre of flotation is at F_1, but if the weight is not too large, it will be approximately over the old centre of flotation. B will move vertically upwards towards the centre of gravity of the added layer, which is somewhere between F and F_1. In this case, both B and G have merely moved upwards and are still in the same vertical line. Consequently, there will be no reason for the ship to change her trim and she will merely sink bodily, increasing her draft by the same amount at each end.

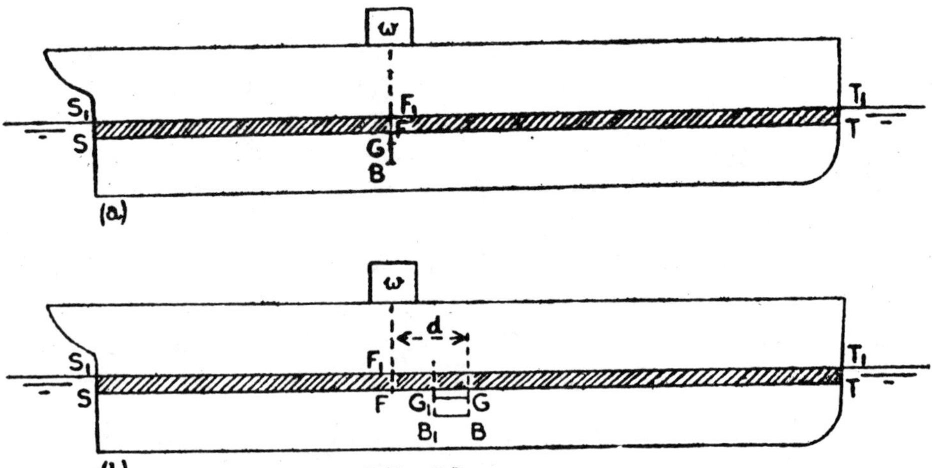

FIG. 72

Now consider what will happen when B and G are not vertically under the centre of flotation, as shown in Fig. 72 (*b*). G will move directly towards the centre of gravity of the weight and B towards the centre of gravity of the added layer, SS_1T_1T; so they both move aft and upwards. Their upward movement will not affect the trim, so we need only consider the fore and aft movement. Thus, we can consider G as moving to G_1 and B to B_1. The horizontal distance between the centre of gravity of the ship and that of the added weight, (d), is the same as that between the centre of buoyancy of the ship and the centre of gravity of the added layer. The shift of B and G can be found in the same way as in transverse stability, that is:—

When W is the weight of the ship; w, the weight added; V the volume of displacement of the ship; and v, the volume of the layer:—

$$GG_1 = \frac{w \times d}{W}$$

$$BB_1 = \frac{v \times d}{V}$$

But the weight of the layer must be equal to the added weight, so:—

$$BB_1 = \frac{w \times d}{W}$$

Therefore $$GG_1 = BB_1$$

Thus B and G will move aft for the same distance and will remain in the same vertical line, so that the ship will again merely sink bodily and will not change her trim.

In either of the cases mentioned above, the sinkage of the ship will be:—

$$\frac{\text{Weight added}}{\text{T.P.C.}}$$

The vessel will increase her draft at each end by the amount of the bodily sinkage

Example 1.—A ship has a T.P.C. of 25·2 tonnes and her drafts are 4·42 metres forward and 4·60 metres aft. What will be her new drafts if a weight of 126 tonnes is added directly over the centre of flotation?

$$\text{Bodily sinkage} = \frac{\text{Weight added}}{\text{T.P.C.}} = \frac{126}{25{\cdot}2} = 5\text{ cm}$$

Original drafts - -	F. 4·42 m	A. 4·64 m
Bodily sinkage - -	+0·05 m	+0·05 m
New drafts - - -	F. 4·47 m	A. 4·69 m

Example 2.—Find the new drafts in a ship which has a T.P.C. of 7·0, if 70 tonnes of cargo are discharged from a point directly below the centre of flotation. The original drafts were 3·05 metres forward and 3·30 metres aft.

$$\text{Bodily rise} = \frac{\text{Weight discharged}}{\text{T.P.C.}} = \frac{70}{7} = 10\text{ cm}$$

Original drafts - -	F. 3·05 m	A. 3·30 m
Bodily rise - - -	−0·10 m	−0·10 m
New drafts - - -	F. 2·95 m	A. 3·20 m

Moderate Weights Loaded off the Centre of Flotation.—When a weight is loaded at any point which is not in the same vertical line as the centre of flotation, the ship will change both her draft and trim. If the weight is of moderate amount the mean draft will only be altered by a few centimetres so that the T.P.C. and the M.C.T.1C. will be nearly the same for both the old and the new drafts. In such a case, we can calculate the new draft and trim as follows:—

(*a*) First, assume the weight to be added at the centre of flotation and calculate the sinkage by the formula:—

$$\text{Bodily sinkage} = \frac{w}{\text{T.P.C.}}$$

(*b*) Next, assume the weight to be shifted from the centre of flotation, forward or aft, to its new position. Calculate the change of trim caused by this by the formula:—

$$\text{Change of trim} = \frac{\text{M.C.T.}}{\text{M.C.T.1C.}}$$

(*c*) Calculate the change of draft at either end, due to the above change of trim, as described in "Change of draft due to change of trim".

(*d*) Add the results of (*a*) to the draft at each end; then apply the results of (*c*). This will give the ship's new drafts after the weight has been added.

(*e*) When weights are discharged, we follow the same procedure as for weights loaded. Remember, however, that they will cause a bodily rise, instead of sinkage: also be careful about which way they will cause the trim to change.

(*f*) If a number of weights are loaded or discharged, use the total effects of all of them. The bodily rise or sinkage will equal the nett weight, divided by the T.P.C. The moment changing trim will be the algebraic sum of all the moments. The following Example 3 illustrates this.

Example 1.—A weight of 200 tonnes is loaded into a ship at a distance of 15 metres abaft amidships. The ship's length is 120 metres, T.P.C. is 10·0, M.C.T.1C. is 96 and the centre of flotation is 3·0 metres abaft amidships. If the original drafts were 5·85 metres forward and 6·00 metres aft, find the new drafts.

$$\text{Bodily sinkage} = \frac{\text{Added weight}}{\text{T.P.C.}} = \frac{200}{10} = 20 \text{ cm}$$

Distance of added weight from F = 15 − 3 = 12 m.

$$\text{Change of trim} = \frac{\text{Moment changing trim}}{\text{M.C.T.1C.}} = \frac{200 \times 12}{96} = 25 \text{ cm}$$

$$\text{Change of draft due to trim: Forward} = \frac{63}{120} \times 25 = 13 \text{ cm}$$

$$\text{Aft} = \frac{57}{120} \times 25 = 12 \text{ cm}$$

Original drafts - -	F. 5·85 m	A. 6·00 m
Bodily sinkage - -	+0·20 m	+0·20 m
	6·05 m	6·20 m
Change due to trim -	−0·13 m	+0·12 m
New drafts - - -	F. 5·92 m	A. 6·32 m

Example 2.—A vessel floats at drafts of 5·12 m forward and 4·69 m aft.

Her T.P.C. is 14·4; M.C.T.1C. is 102; and F is amidships. What will be the new drafts after 98 tonnes of water has been pumped out of the fore peak tank, 52 m forward of amidships?

$$\text{Bodily rise} = \frac{\text{Weight}}{\text{T.P.C.}} = \frac{98}{14 \cdot 4} = 7\text{cm}$$

$$\text{Change of trim} = \frac{\text{M.C.T.}}{\text{M.C.T.1C.}} = \frac{98 \times 52}{102} = 50\text{cm}$$

Since F is amidships, the drafts fore and aft will each change by half the change of trim.

Old drafts	F. 5·12m	A. 4·69m
Bodily rise	— ·07m	— ·07m
	5·05m	4·62m
½ change of trim	— ·25m	+ ·25m
New drafts	F. 4·80m	A. 4·87m

Example 3.—A ship is 140 metres long, has a T.P.C. of 20 and an M.C.T.1C. of 120, whilst the centre of flotation is 3·0 metres abaft amidships. The draft is 7·10 metres forward and 7·25 metres aft. The following cargo is then worked:—

(*a*) Loaded - 120 tonnes, 50 metres forward of F.
(*b*) Loaded - - 70 ,, 20 ,, abaft F.
(*c*) Discharged - 90 ,, 30 ,, forward of F.

Find the new drafts.

Weight	Distance from F	Moment
120 tonnes (+)	50 m. (Forward)	6000 by head
70 ,, (+)	20 m. (Aft)	1400 by stern
90 ,, (−)	30 m. (Forward)	2700 by stern
100 tonnes (+)		1900 by head

$$\text{Bodily sinkage} = \frac{\text{Nett weight loaded}}{\text{T.P.C.}} = \frac{100}{20} = 5 \text{ cm}$$

$$\text{Change of trim} = \frac{\text{Moment changing trim}}{\text{M.C.T.1C.}} = \frac{1900}{120} = 16 \text{ cm}$$

$$\text{Change of draft for trim: Forward} = \frac{73}{140} \times 16 = 8 \text{ cm}$$

$$\text{Aft} = \frac{67}{140} \times 16 = 8 \text{ cm}$$

Original drafts - -	F. 7·10 m	A. 7·25 m
Bodily sinkage - -	+0·05 m	+0·05 m
	7·15 m	7·30 m
Change due to trim -	+0·08 m	−0·08 m
New drafts - - -	F. 7·23 m	A. 7·22 m

Large Weights Added Off the Centre of Flotation.—When large weights are added to a ship, the methods described in the last section must be modified. This is necessary because, as the ship changes her draft, there may be considerable changes in the position of the centre of flotation and in the amounts of the tonnes per centimetre immersion and moment to change trim by one centimetre. If the weight is not too large (say, not more than 8% or 10% of the ship's deadweight), the following method will give results which are accurate enough for most practical purposes. For larger weights, or where greater accuracy is required, it is better to use moments about the after perpendicular, as described later in this chapter.

(*a*) Find the approximate bodily sinkage, as shewn in (*a*) above, using the original tonnes per centimetre immersion. Apply this sinkage to the old draft to find an approximate new mean draft and find the tonnes per centimetre immersion for this new draft. Find the mean of the two tonnes per centimetre immersion by adding them together and dividing by two. Use this mean T.P.C. to find the true sinkage and new mean draft, using the formula:—

$$\frac{w}{\text{T.P.C.}}$$

Alternatively, if a sufficiently accurate displacement curve or scale is available, this may be used to find the new draft, as shewn in the following Example 2.

(*b*) Find the position of the centre of flotation for this new mean draft, using the ship's stability curves. Also the moment to change trim by one centimetre at this draft. Use these to find the change of trim by the formula:—

$$\text{Change of trim} = \frac{\text{M.C.T.}}{\text{M.C.T.1C.}}$$

(*c*) Calculate the change of draft due to the change of trim, using the new position of the centre of flotation.

(*d*) Find the new drafts fore and aft by applying sinkage and changes, as in the last section.

Example 1.—An oil tanker, which is light, is found to be trimmed too much by the stern. In order to change the trim and also to increase the draft, a tank which has its centre of gravity 70 metres abaft the stem is filled with 1100 tonnes of water. The following details are found from the stability scales:—

Draft	T.P.C.	M.C.T.1C.	Centre of Flotation
4·00 m	20·8	160	2·9 m abaft amidships
4·50 m	21·2	165	3·0 m abaft amidships
5·00 m	21·7	170	3·1 m abaft amidships

If the original drafts were 2·80 metres forward and 5·20 metres aft, find the new drafts. The ship is 200 metres long.

The original mean draft of 4·00 m gives a T.P.C. of 20·8.

$$\text{Approximate sinkage} = \frac{\text{Added weight}}{\text{T.P.C.}} = \frac{1100}{20 \cdot 8} = 53 \text{ cm}$$

Approximate new mean draft = 4·00 + 0·53 = 4·53 m
T.P.C. for draft of 4·00 m = 20·8
T.P.C. for draft of 4·53 m = 21·2

Mean T.P.C. = 21·0

$$\text{Corrected sinkage} = \frac{\text{Added weight}}{\text{T.P.C.}} = \frac{1100}{21 \cdot 0} = 52 \text{ cm}$$

Corrected new mean draft = 4·00 + 0·52 = 4·52 m

At this draft, the M.C.T.1C. is 165 and the centre of flotation is 3·0 metres abaft amidships.

The tank will be 33 metres forward of the centre of flotation.

$$\text{Change of trim} = \frac{\text{Moment changing trim}}{\text{M.C.T.1C.}} = \frac{1100 \times 33}{165} = 220 \text{ cm}$$

$$\text{Change of draft for trim: Forward} = \frac{103}{200} \times 220 = 113 \text{ cm}$$

$$\text{Aft} = \frac{97}{200} \times 220 = 107 \text{ cm}$$

Original drafts - -	F. 2·80 m	A. 5·20 m
Bodily sinkage - -	+0·52 m	+0·52 m
	3·32 m	5·72 m
Change for trim - -	+1·13 m	−1·07 m
New drafts - - -	F. 4·45 m	A. 4·65 m

Example 2.—The ship for which the hydrostatic particulars are given in the back of this book, is 140 m long and floats at drafts of 5·56 m forward and 6·00 m aft. She then loads 600 tonnes of cargo at 35 m forward of *F*; 1800 tonnes at 2 m forward of *F*; and 1400 tonnes at 31 m abaft *F*. Find her new drafts.

Present mean draft is 5·58 m

The hydrostatic particulars shew that *F* is amidships at this draft, so the draft at *F* will also be 5·58 m. The particulars also shew that the displacement at this draft is 9702 tonnes.

Weight	Distance from *F*	Moment
600 t	35 m (F)	21000 by head
1800 t	2 m (F)	3600 by head
1400 t	31 m (A)	43400 by stern
3800 t		18800 by stern

New displacement = 9702 + 3800 = 13502 tonnes
New draft at *F* for 13502 t displacement = 7·32 m
M.C.T.1C. at draft of 7·32m = 196·9 t/m

$$\text{F} \quad ,, \quad ,, \quad ,, \quad ,, \quad = \begin{cases} 68{\cdot}3\text{m from aft} \\ 1{\cdot}7\text{m abaft amidships} \end{cases}$$

$$\text{Change of trim} = \frac{\text{M.C.T.}}{\text{M.C.T.1C.}} = \frac{18800}{196{\cdot}9} = 95\text{cm (by stern)}$$

But, the ship was originally trimmed 4 cm by stern
So, the new trim = 95 + 4 = 99 cm by stern

$$\text{Difference of draft for trim: Forward} = \frac{71{\cdot}7}{140} \times 99 = 51\text{cm}$$

$$\text{Aft} = \frac{68{\cdot}3}{140} \times 99 = 48\text{cm}$$

New draft at F.	F. 7·32m	A. 7·32m
Difference for trim	— ·51m	·48m
New Drafts	F. 6·81m	A. 7·80m

Special Trim or Draft.—It is often desirable to obtain a certain trim when the ship is loaded. Generally, we do not require the trim to exact limits, but merely "a little by the stern". In this case, it is usually sufficient to load weights as desired during the earlier stages of loading. Later on, we can get the trim roughly as we want it by using a combination of experience and common sense. Finally, near the end of the loading, we can trim our ship exactly as we want her by "juggling" with the M.C.T.1C. and the amount of weight to come on board.

Sometimes, as when a ship has to cross a bar, we require a certain maximum draft, or wish to keep the after draft constant. This can be done quite easily.

Loading Weights to Obtain a Desired Trim.—First, find the difference of the drafts forward and aft, thus finding the existing trim. The difference between this and the desired trim is the change of trim required.

Let t be the existing trim and t_1 the desired trim. Let w be the added weight and d its distance from the centre of flotation.

$$\text{Change of trim required} = t \sim t_1$$

but,

$$\text{Change of trim} = \frac{\text{M.C.T.}}{\text{M.C.T.1C.}}$$

so,

$$\frac{\text{M.C.T.}}{\text{M.C.T.1C.}} = t \sim t_1$$

$$\text{M.C.T.} = \text{M.C.T.1C.}\ (t \sim t_1)$$

Thus, according to which we wish to find:—

$$d = \frac{\text{M.C.T.1C.}\ (t \sim t_1)}{w}$$

$$w = \frac{\text{M.C.T.1C.}\ (t \sim t_1)}{d}$$

Example 1.—A ship has an M.C.T.1C. of 150 tonne-metres and floats at drafts of 6·00 metres forward and 7·00 metres aft. How much water must be run into a tank, the centre of gravity of which is 50 metres forward of the centre of flotation, to bring the ship to a trim of 20 centimetres by the stern?

$$\text{Present trim } (t) = 100 \text{ cm by stern}$$

$$\text{Required trim } (t_1) = 20 \text{ cm by stern}$$

$$(t \sim t_1) = 80 \text{ cm by head}$$

Let w be the required weight

Then:

$$w \times d = \text{M.C.T.1C.}\ (t \sim t_1)$$

$$50w = 150 \times 80$$

$$w = \frac{150 \times 80}{50} = 240 \text{ tonnes.}$$

Example 2.—A ship, which is completing loading, has drafts of 7·58m forward and 7·72 m aft. Her M.C.T.1C. is 118 t/m. 360 tonnes of cargo remain to be loaded and this is to be distributed between No. 1 hold (45 m forward of F) and No. 4 hold (25 m abaft F). How much cargo should there be loaded in each hold in order to finish with a trim of 50 cm by the stern?

Present trim = 7·72 − 7·58 = 14cm by stern
Required trim 50cm by stern

Required change of trim 36cm by stern

Load w tonnes into No. 1; $(360 - w)$ tonnes into No. 4.

Moment of No. 1 = $45w$ t/m by head.

Moment of No. 4 = $25(360 - w)$ t/m by stern.

To increase the trim by the stern, the moment of No. 4 must be greater than that of No. 1.

So,
$$\begin{aligned} \text{M.C.T.} &= \text{Moment of No. 4} - \text{Moment of No. 1} \\ &= 25(360 - w) - 45w \\ &= 9000 - 70w \end{aligned}$$

$$\text{Change of trim} = \frac{\text{M.C.T.}}{\text{M.C.T.1C.}}$$

$$36 = \frac{9000 - 70w}{118}$$

$$w = \frac{9000 - 36 \times 118}{70} = 68 \text{ tonnes in No. 1}$$

$$\begin{array}{r} 360 \\ \hline 292 \text{ tonnes in No. 4} \end{array}$$

Position to Load Weight so as not to Change the Draft Aft.—When a weight is loaded, there is usually a bodily sinkage of the ship and also a change of trim. The bodily sinkage tries to increase the draft aft: whilst if the weight is loaded forward of the centre of flotation, the change of trim tries to decrease the after draft. It is possible by balancing these effects against each other, to keep the after draft constant when a weight is loaded.

Let w be the added weight and d its distance forward of the centre of flotation. Let L be the length of the ship and l the distance of the centre of flotation from aft.

$$\text{Bodily sinkage} = \frac{\text{T.P.C.}}{w}$$

This is equivalent to an increase of draft aft.

$$\text{Total change of trim} = \frac{w \times d}{\text{M.C.T.1C.}}$$

$$\text{Change of draft aft, due to trim} = \frac{w \times d}{\text{M.C.T.1C.}} \times \frac{l}{L}$$

This is equivalent to a decrease of draft aft.

For the after draft to remain constant, the increase due to the sinkage must be equal to the decrease due to the change of trim. That is:—

Bodily sinkage = Change of draft aft, due to trim.

$$\frac{w \times d}{\text{M.C.T.1C.}} \times \frac{l}{L} = \frac{w}{\text{T.P.C.}}$$

$$\frac{w \times d}{\text{M.C.T.1C.}} = \frac{w \times L}{\text{T.P.C.} \times l}$$

$$d = \frac{w \times L \times \text{M.C.T.1C.}}{w \times \text{T.P.C.} \times l}$$

$$d = \frac{\text{M.C.T.1C.} \times L}{\text{T.P.C.} \times l}$$

Note that the after draft will remain constant if the weight is loaded in the position so found, irrespective of the amount of weight added or discharged, within reasonable limits.

Example.—A ship which is loading cargo has drafts of 6·50 metres forward and 7·00 metres aft, when there are a further 200 tonnes of cargo to come on board. Where must this cargo be placed in order to maintain this draft of 7·00 metres aft? The ship is 140 metres long, her T.P.C. is 20, M.C.T.1C. is 135 and the centre of flotation is 3·0 metres abaft amidships.

Distance of F from aft = 70 − 3 = 67 m.

$$d = \frac{\text{M.C.T.1C.} \times L}{\text{T.P.C.} \times l} = \frac{135 \times 140}{20 \times 67} = 14{\cdot}1 \text{ m forward of F}$$

Load cargo 14·1 metres forward of F; or 11·1 metres forward of amidships.

Loading a Weight to Produce a Desired Draft Aft.—This may be achieved by a little modification of the above. When a weight is loaded, the bodily sinkage causes an increase of draft aft, whilst the change of trim may cause an increase or decrease, according to whether the weight is loaded forward of or abaft the centre of flotation. So:—

To obtain a decrease of draft, aft:—

Change of draft aft = change due to trim − bodily sinkage.

To obtain an increase of draft, aft:—

Change of draft aft = Bodily sinkage ± change due to trim.

In the latter case, if the bodily sinkage is less than the change required, the sign will be + and the weight loaded abaft the centre of flotation. If it is greater, the sign will be − and the weight loaded forward of the centre of flotation.

Example.—A ship has a draft, aft, of 5·14 m. She is 120 m long, has a T.P.C. of 15·1, M.C.T.1C. of 102, and F is 2·0 m forward of amidships. Where must 90 tonnes of cargo be loaded in order that the ship may sail with a draft, aft, of 5·00 m?

Required decrease of draft, aft $= 5{\cdot}14 - 5{\cdot}00 =$	14cm
Bodily sinkage $= \dfrac{w}{\text{T.P.C.}} = \dfrac{90}{15{\cdot}1}$	6cm
Required rise, aft, due to trim	20cm

To cause the ship to rise, aft, we must load the weight forward of F.

$$\text{Rise, aft, due to trim} = \frac{w \times d}{\text{M.C.T.1C.}} \times \frac{62}{120}$$

$$20 = \frac{90d}{102} \times \frac{62}{120}$$

$$d = \frac{20 \times 102 \times 120}{90 \times 62} = 43{\cdot}9\text{m (forward of } F)$$

Maximum Weight to Load for a Given Draft.—We may wish to load as much cargo as possible, but to sail with a given maximum draft in order to cross a bar or a dock sill. This can be done by bringing the ship on to an even keel at the required draft.

Example.—A ship has drafts of 6·72 m forward and 6·94 m aft. Her T.P.C. is 21·7, M.C.T.1C. is 183, and F is 1·5 m abaft amidships. In order to cross a bar, her draft on sailing must not exceed 7·00 m. What is the maximum weight of cargo that she can load, and where must it be placed?

Bring the ship to an even keel at 7·00 m draft.

Present mean draft	6·83m	
Maximum permissible draft	7·00m	
Maximum sinkage	·17m	(17cm)

$$\text{Weight to load} = \text{Sinkage} \times \text{T.P.C.} = 17 \times 21{\cdot}7 = 369\text{t}$$

$$\text{Present trim} = 6{\cdot}94 - 6{\cdot}72 = 22\text{cm by stern}$$

To bring the ship to an even keel, the cargo must be loaded forward of F so as to change the trim by 22cm by the head.

$$\text{Change of trim} = \frac{w \times d}{\text{M.C.T.1C.}}$$

$$22 = \frac{369d}{183}$$

$$d = \frac{22 \times 183}{369} = 10{\cdot}9\text{m forward of } F$$

The Use of Moments About the After Perpendicular.—This is an alternative method of finding moment to change trim when weights are loaded or discharged. It simplifies the working, in that the moments of all weights loaded are added, whilst the moments of all weights discharged are subtracted: so that it is not necessary to consider the effect on the trim of each separate weight as we go along.

To use the method, we must be able to find:—

(*a*) the distance of the centre of buoyancy from the after perpendicular at any draft;

(*b*) The distance of each parcel of cargo from the after perpendicular.

The Stability Information Booklet, as recommended by the Department of Trade, gives the above information.

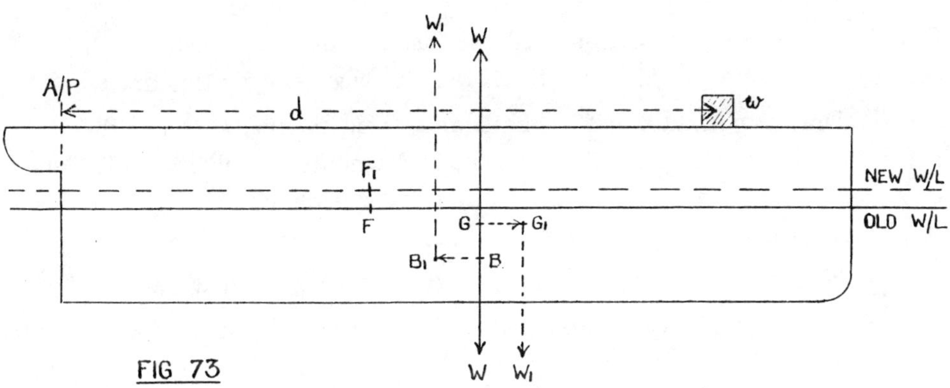

FIG 73

Fig. 73 illustrates the method. Before any weights are loaded, B and G must be in the same vertical line in order that the ship may be in equilibrium, fore and aft. Let the ship's displacement be W tonnes and let B and G be at a distance of x metres from the after perpendicular.

The moment of gravity about the A.P. $= Wx$ tonne-metres.

Now let a weight of w tonnes be loaded at a distance of d metres from the after perpendicular. This will cause the ship to sink bodily to some new mean draft. It will also cause G to move to G_1, so that the new displacement (W_1) will now act vertically down through G_1. The ships' moment of gravity about the after perpendicular will now be W_1 multiplied by the distance of G_1 from the after perpendicular.

If the weight loaded is large, the bodily sinkage may cause B to move a little: say, to some point B_1. We must, therefore, use the stability information to find the distance of B_1 from the after perpendicular. The new displacement

W_1 will now act vertically upward through B_1: so the moment of buoyancy about the after perpendicular will now be W_1 multiplied by the distance of B_1 from the after perpendicular.

We now have the moment of gravity acting downwards through G_1 and the moment of buoyancy acting upwards through B_1. The difference between the two will be the moment trying to change the ship's trim. That is:—

Moment to change trim = Moment of gravity $\sim$ Moment of buoyancy.

If the moment of gravity is the greater, the ship will now trim by the head. If the moment of buoyancy is the greater, the ship will trim by the stern.

To use this method, therefore, we must:—

(*a*) If necessary, correct the original mean draft to give the draft at F. From the stability information, find the distance of B (and hence of G) from the after perpendicular and also the ship's displacement for this draft.

(*b*) Draw up a table of weights and their moments about the after perpendicular, including those of the ship. Moments of weights loaded must be added, those of weights discharged must be subtracted. This will give the ship's new displacement and moment of gravity.

(*c*) Find the bodily rise or sinkage and, from this, the new draft at F.

(*d*) Use the stability information to find the new distance of B from the after perpendicular for the new draft. This, multiplied by the new displacement, will give the new moment of buoyancy.

(*e*) The difference between the new moments of gravity and buoyancy will be the moment to change the trim. If the moment of gravity is the greater, the ship will trim by the head: if the moment of buoyancy is the greater, she will trim by the stern.

(*f*) We can now use the methods described previously to find the new trim and drafts fore and aft.

Example 1.—A ship has a displacement of 8600 tonnes at a mean draft of 5·01 m. B is 71·90 m from the after perpendicular and the T.P.C. is 20·60 tonnes. She then loads 600 t at 40 m from the after perpendicular, 300 t at 110 m, and 400 t at 70 m. She also discharges 500 t from 50 m from the after perpendicular, and 400 t from 60 m. The stability information then shews that at a mean draft of 5·20 m, B is 71·80 m from the after perpendicular and the T.P.C. is still 20·60 t. Find the moment changing the trim.

	Weight	d from A/P	Moment about A/P
Ship	8600	71·90	618340
Loaded	+ 600	40	+ 24000
,,	+ 300	110	+ 33000
,,	+ 400	70	+ 28000
Discharged	— 500	50	— 25000
,,	— 400	60	— 24000
	9000		654340

Added weight = 9000 — 8600 = 400 tonnes

$$\text{Bodily sinkage} = \frac{w}{\text{T.P.C.}} = \frac{400}{20\cdot6} = 19\cdot4\text{cm}$$

New mean draft = 5·01 + 0·19 = 5·20m

Moment of buoyancy = Distance of B from A/P × new W
= 71·80 × 9000 = 646200 t/m

Moment changing trim = moment of gravity $\sim$ moment of buoyancy
654340 $\sim$ 646200
8140 t/m by the head.

Example 2.—The hydrostatic particulars given in the back of this book are for a vessel of 140m long. She is floating at drafts of 6·14m forward and 6·26 m aft. She then loads 800 t at 52 m from the after perpendicular, 700 t at 74 m; and discharges 550 t from 80 m. Find the new drafts?

Old mean draft = 6·20m

At this draft, F is 0·67m abaft amidships

Correction to draft at F = 0·67 × 0·12 ÷ 140 = ·0006m

So, corrected draft at F = 6·20m

For draft of 6·20 m, the displacement is 11027 t and B is 71·53 m from the after perpendicular.

	Weight	d from A/P	Moment about A/P
Ship	11027	71·53	788761
Loaded	+ 800	52	+ 41600
,,	+ 700	74	+ 51800
Discharged	— 550	80	— 44000
	11977		838161

From the hydrostatic particulars, with new displacement of 11977 tonnes:—

$$\begin{aligned}\text{New draft at } F &= 6{\cdot}64\text{m}\\ \text{M.C.T.1C.} &= 185{\cdot}4 \text{ t/m}\\ \text{LCB} &= 71{\cdot}30\text{m from A/P}\\ \text{LCF} &= \begin{cases}68{\cdot}80\text{m from A/P}\\ 71{\cdot}20\text{m from F/P}\end{cases}\end{aligned}$$

New moment of buoyancy = 11977 × 71·30 =	853960
New moment of gravity	838161
M.C.T. = Excess moment of buoyancy	15799

$$\text{Change of trim} = \frac{\text{M.C.T.}}{\text{M.C.T.1C.}} = \frac{15799}{185{\cdot}4} = 85\text{cm by stern}$$

$$\text{Original trim} = 12\text{cm by stern}$$

$$\text{New trim} = 97\text{cm by stern}$$

$$\text{Change from even keel drafts: Forward} = \frac{71{\cdot}2}{140} \times 97 = 49\text{cm}$$

$$\text{Aft} = \frac{68{\cdot}8}{140} \times 97 = 48\text{cm}$$

New even keel drafts	F. 6·64m	A. 6·64m
Difference for trim	− 0·49m	+ 0·48m
New drafts	F. 6·15m	A. 7·12m

CHAPTER 13

STABILITY CURVES AND SCALES

When a ship is built, the Naval Architects calculate certain data affecting her stability and set it out in the form of curves and scales. Some of this information must be supplied to ships, as described in Chapter 15. Meanwhile, let us consider the basic ones and their uses.

Hydrostatic Curves.—In the back of this book will be found a set of curves of a type usually supplied to ships. As we have already mentioned, such curves vary considerably in detail, but are fundamentally the same, so that anyone who can understand those given here should have no difficulty in taking off information from any others he may encounter.

A scale of mean drafts of the ship runs vertically up the left-hand side of the sheet, lines being drawn horizontally across the plan at every half-metre. The main body of the plan consists of a number of curves, each shewing the amount or position of any one item for any mean draft on the scale. Along the top or bottom edges are scales from which the amounts of the various items can be read off.

To obtain information, find the ship's mean draft on the left-hand scale and draw a line horizontally across the plan from this, until it cuts the curve required; or use dividers to measure up to this point from the nearest horizontal line. From the point thus found, drop a perpendicular line to the appropriate scale on the bottom of the plan, or again use dividers, and read off the required information.

It will be noticed that the curves have their descriptions written on them in full, whilst, to save space, abbreviations have been used for naming the scales. Sometimes abbreviations are used along the curves also, but this should not cause any difficulty, since they are standardised and should be known by anyone studying stability. Those used in the curves given here are as follows:—

Height of the centre of buoyancy above the keel	KB
Height of the transverse metacentre above the keel ..	KM
Height of the longitudinal metacentre above the keel.. ..	KM_L
Moment to change trim by one centimetre	M.C.T.1C.
Tonnes per centimetre immersion	T.P.C.
Centre of buoyancy from A.P.	B. from A.P.
Centre of flotation from A.P.	F. from A.P.
Displacement	"Displacement."

Always be careful that you take off information from the proper scale.

Use of the Hydrostatic Curves.—The uses of the information which we can obtain from the curves are fairly obvious to anyone who has read through this book. Let us consider an example of the information that it is possible to obtain.

Suppose that we wish to find all possible stability information from the curves given in the back of this book, assuming that the ship is floating at a mean draft of 5·40 metres. We draw a horizontal line across the plan at this draft and can then take off the following information from the scales at the foot:—

Displacement	9300 tonnes
KB	2·93 metres
KM	8·18 metres
KM_L	247 metres
M.C.T.1C.	162 tonne-metres
T.P.C.	20·9 tonnes
Centre of flotation from A.P.	70·2 metres
Centre of buoyancy from A.P.	71·9 metres

The Deadweight Scale.—This scale is familiar to most ship's officers and is another method of giving certain stability information which they are most likely to need. A typical scale will be found in the back of this book, with the hydrostatic curves, and is made out for the same ship as the latter.

To obtain information from the scale, draw a horizontal line, or lay a ruler, across the scale at the appropriate draft, then read-off the figures shewn against this.

For example, to obtain information from the scale for a draft of 6·40m, lay a ruler across it at that draft and you will find the following:—

Deadweight in salt water	6900 tonnes
Displacement in salt water	11450 tonnes
T.P.C. in salt water	21·8 tonnes
M.C.T.1C. in salt water	181 tonne-metres
Deadweight in fresh water	6600 tonnes
Displacement in fresh water	11150 tonnes
T.P.C. in fresh water	21·3 tonnes

Hydrostatic Particulars.—Sometimes, instead of hydrostatic curves, similar information is set out in tabular form, as "Hydrostatic Particulars". This is the method recommended for the Stability Information Booklet which is described in Chapter 15. Its main advantage is that it will usually give more accurate information than that which we could obtain from the hydrostatic curves or the deadweight scale. An example of such a table, in the recommended form, is given at the back of this book.

To obtain information from the table for a certain draft, take it out for the nearest draft, or drafts and then interpolate, as necessary, according to the accuracy required. For instance, suppose that we wish to find information for a draft of 6·34 m: we take out the information for drafts of 6·20 m and 6·40 m and then interpolate between these for the required draft. This will give:—

Displacement	11330 tonnes
T.P.C.	21·75 tonnes
M.C.T.1C.	180·09 tonne-metres
LCB from the After Perpendicular	71·50 metres
LCF from the After Perpendicular	69·20 metres
VCB above base	3·45 metres
KM	8·03 metres
KM_L	226·2 metres

Curves of Statical Stability.—These are graphs which show the ship's statical stability at all angles of heel Each curve is given for one displacement only, using an assumed *KG*. The curves sometimes show righting moments ($W \times GZ$), or sometimes righting levers (*GZ*), according to the purpose for which they are intended.

There is a theorem which says that we can find the shape of the curve at its origin (*i.e.*, 0°) by drawing a perpendicular at 1 radian (57·3°) and on this, marking off the *GM* from the *GZ* scale. A straight line from the point so found to the point of origin will form a tangent to the curve at that point.

Figure 74 shows a single curve, on which the righting levers have been plotted. It has been constructed by calculating the *GZ*'s at various angles of heel and plotting these as a graph. The dotted lines illustrate the method of finding the shape of the curve at its origin, referred to in the theorem above: the assumed *GM* has been marked off on the perpendicular at 57·3° and the line from here to the point of origin forms the tangent to the curve.

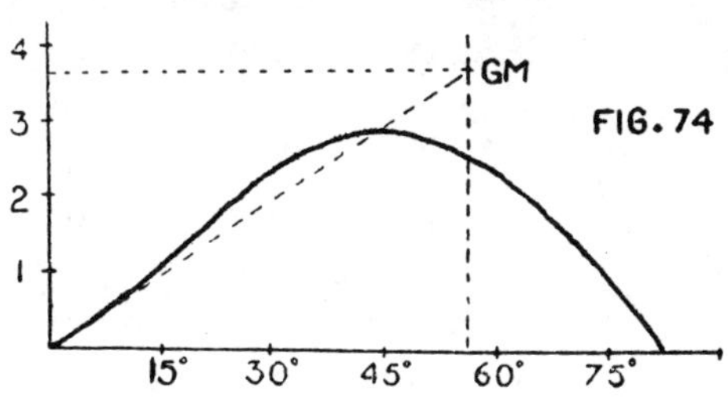

FIG. 74

A single curve of statical stability enables us to find the range of stability, or angle of vanishing stability; the angle and amount of maximum stability; the approximate *GM* for which the curve has been drawn. The latter is found by drawing a tangent to the curve at the point of origin and then measuring off the *GM* at the point where this tangent meets the perpendicular at 1 radian.

FIG. 75

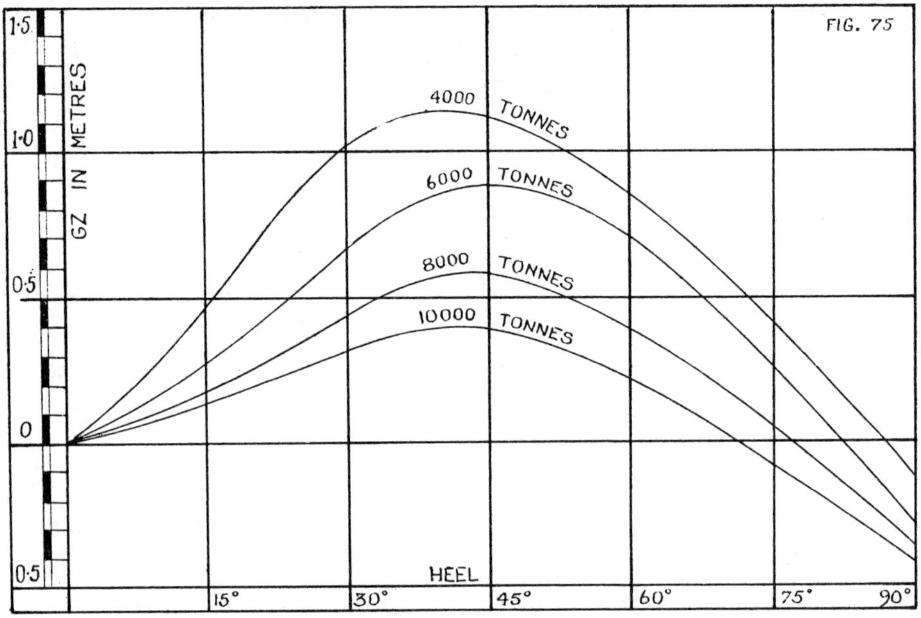

If we wish for a comprehensive picture of the stability of a ship we may construct a series of curves of statical stability, using the same *KG* for all, but a different displacement for each. A typical set of such curves, drawn for one ship, is shewn in Figure 75, above.

Cross Curves.—These are another common method of drawing graphs of righting levers. They are different from the ordinary curves of statical stability, however, because each curve shows the righting levers for one angle of heel only, but at different displacements. The principle on which they are constructed is shewn diagrammatically in Figure 76; in which the curves of statical stability are represented as standing-up, one behind the other, at equal distances apart Points 15° apart have

FIG. 76

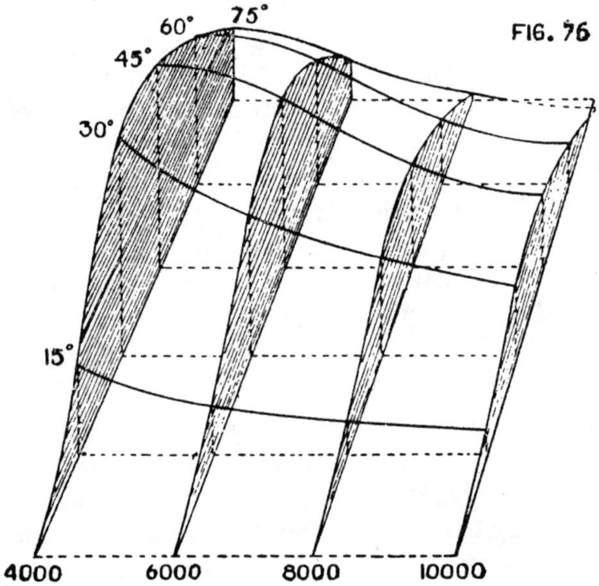

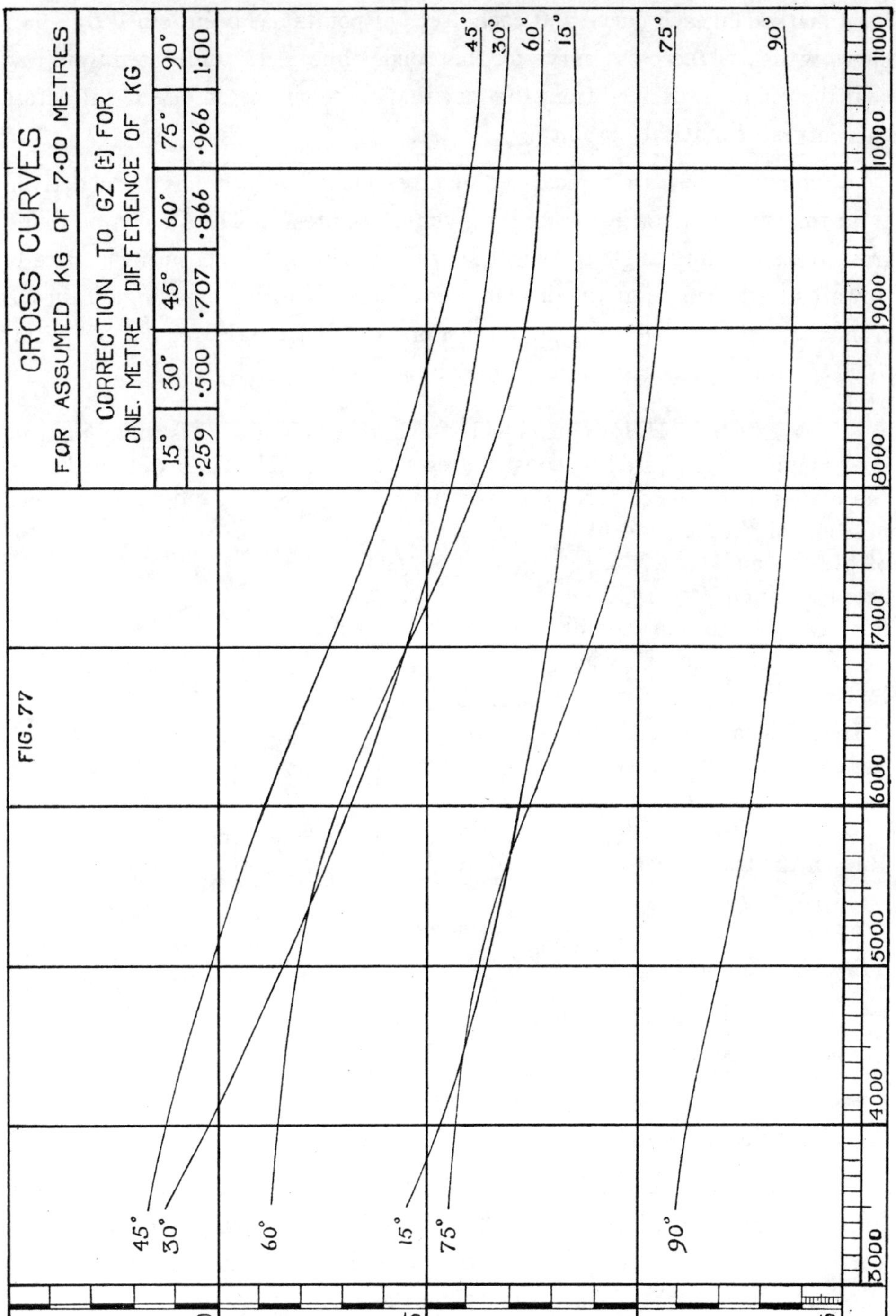
CROSS CURVES
FOR ASSUMED KG OF 7·00 METRES
CORRECTION TO GZ [±] FOR
ONE. METRE DIFFERENCE OF KG
15° 30° 45° 60° 75° 90°
·259 ·500 ·707 ·866 ·966 1·00
1·0
0·5
0
0·5
3000 4000 5000 6000 7000 8000 9000 10000 11000
45° 30° 60° 15° 75° 90°

FIG. 77

been marked on each curve and each "set" of points has been joined by a fair curve, which is the cross curve for that angle of heel. It can be seen from this that their name is derived from the fact that these curves, in effect, cut across the curves of statical stability.

Figure 77 shews a typical set of cross curves, which has been derived from the curves of statical stability given in Figure 75. The GZ can be found from these, for any angle for which a curve is given, by measuring vertically upwards to the curve, at the displacement chosen, and then reading off the GZ from the scale on the left-hand side. For example, the GZ for 30° of heel at 7000 tonnes displacement would be 0·55 metres.

The Effect of Height of G.—If the actual height of G is different from the assumed height used in constructing the curves, the GZ's obtained from the curves will be incorrect. For instance, in Fig. 78, let G be the assumed position of the ship's centre of gravity and let G_1 be its actual position. The righting lever given by the curves will be GZ, but the true righting lever will be G_1Z_1.

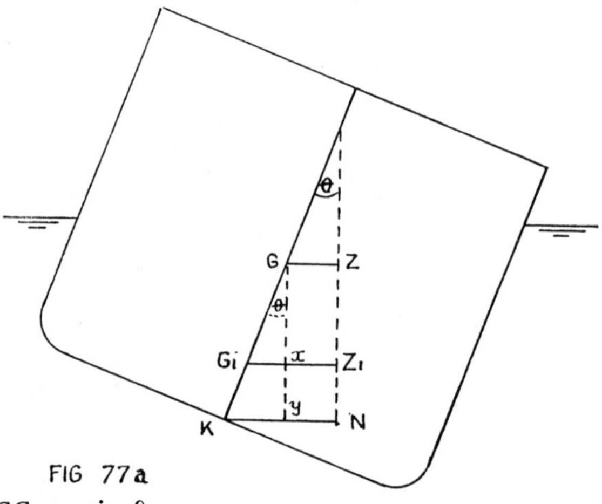

FIG 77a

If Gx is a vertical line, parallel to ZZ_1, then:—

$$GZ = xZ_1$$

$$G_1Z_1 = GZ + G_1x$$

But, in the triangle GG_1x:—

$$G_1x = GG_1 \times \text{Sin } \theta$$

Thus, in this case,

True GZ = GZ from curves + $GG_1 \times \sin\theta$.

On the other hand, if G were the true position and G_1 the assumed position of the ship's centre of gravity:—

$$GZ = G_1Z - G_1x$$

And True GZ = GZ from curves — $GG_1 \times \text{Sin } \theta$

So, if the actual KG is different from that used in the curves, we must apply a correction of $GG_1 \times \text{Sin } \theta$ to the GZ found from the curves. This correction must be added if the true KG is less than the assumed KG: or subtracted if the true KG is greater than the assumed KG.

For convenience, a table of corrections is often placed on a set of cross curves, as in the right-hand top corner of Fig. 77. This is actually a table of natural sines; which gives us Sin θ, for use as above.

Example.—The cross curves shew a GZ of 0·73 m at 45° heel and displacement 7000 tonnes, for an assumed KG of 7·00 m. If the actual KG is 7·54 m, what is the true GZ at 45°?

$$\text{Correction} = GG_1 \times \text{Sin } \theta = 0{\cdot}54 \times \text{Sin } 45^\circ$$
$$= 0{\cdot}38\text{m}$$

Since the actual KG is greater than the assumed KG, this correction must be subtracted from the GZ found from the curves.

So, $$\text{True } GZ = 0{\cdot}73 - 0{\cdot}38 = 0{\cdot}35 \text{ metres.}$$

KN Curves.—This is another type of cross curve, in which G is assumed to be at the ship's keel (K). If we refer to Fig. 77 (a), we can see that, in this case, KN would be the righting lever. If, however, G were the true position of the ship's centre of gravity, then:—

$$\text{True } GZ = KN - Ky$$
$$= KN - KG \times \sin \theta$$

A typical set of KN curves is shewn in Fig. 78.

The advantage of this method is that the correction is always $KG \times \sin \theta$ and is always subtracted from the KN found from the curves. This reduces the possibility of error.

Example.—The curves shew a KN of 3·98 m at 30° of heel, for a displacement of 9000 tonnes. If the ship's KG is 6·84 m, what is the GZ at 30°?

$$GZ = KN - KG \times \text{Sin } \theta$$
$$= 3{\cdot}98 - 6{\cdot}84 \times \text{Sin } 30^\circ$$
$$= 0{\cdot}56 \text{ metres.}$$

The Metacentric Diagram.—This diagram is another method of showing the heights of the metacentre and centre of buoyancy for various mean drafts. A scale of mean drafts is drawn up and a line is drawn across the paper at an angle of 45° to it—this is called "the 45° line". A curve of centres of buoyancy is drawn in below this line, so that the distance from the curve to

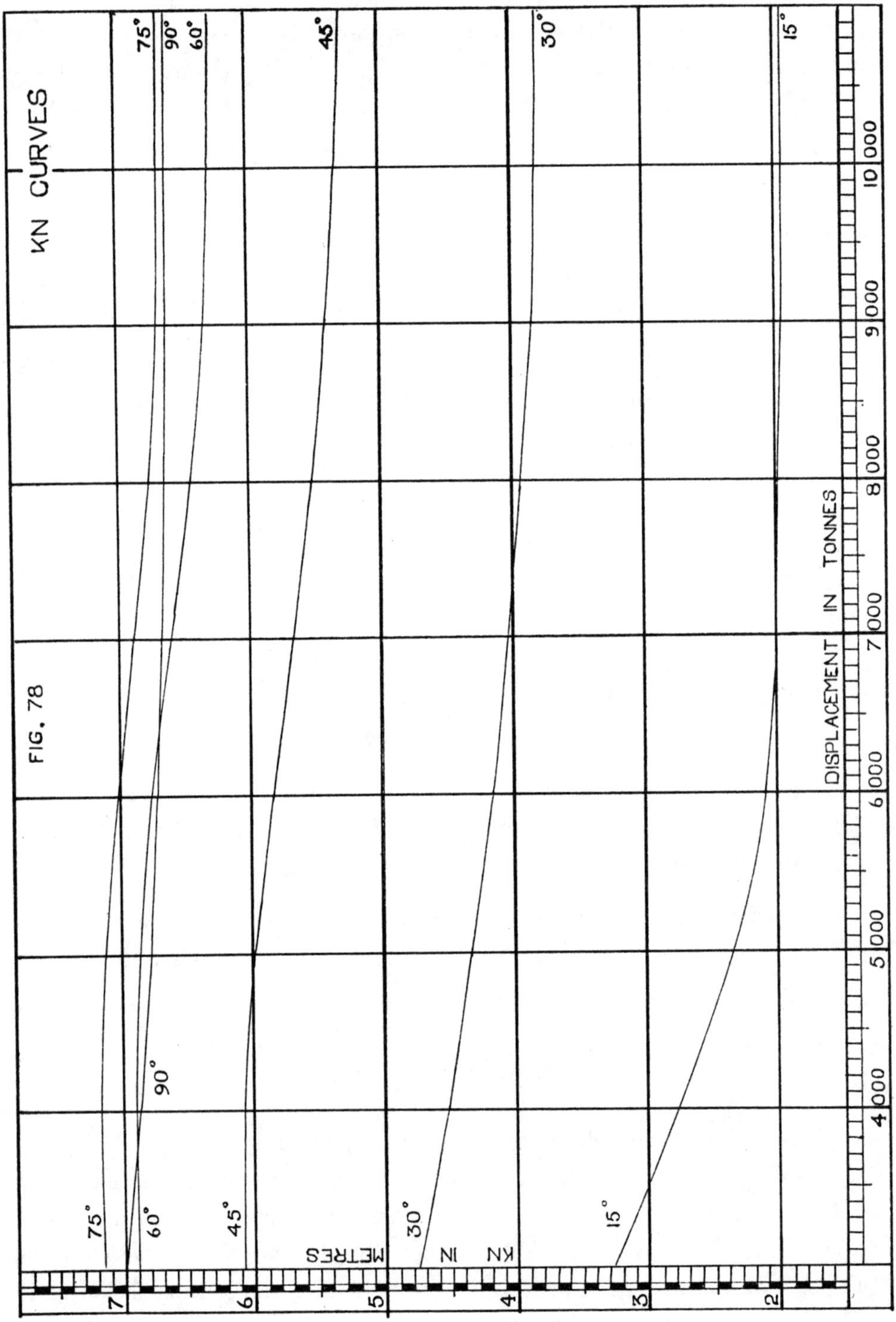

FIG. 78

any point on the line is equal to the distance of the centre of buoyancy below the water line at that point. The curve of metacentres is drawn in a similar way above the 45° line.

FIG. 79

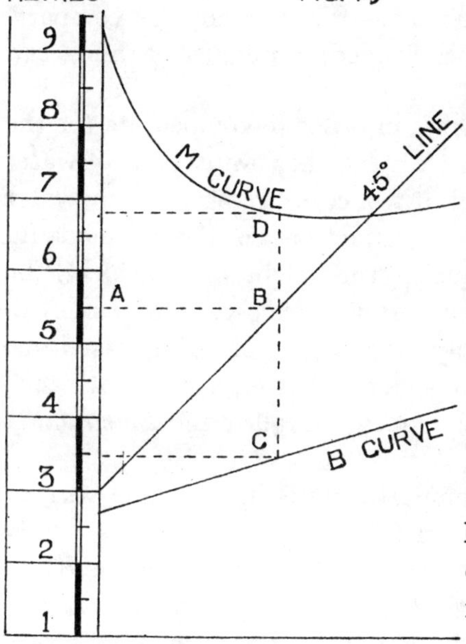

This is best seen from Fig. 79. To find the positions of B and M for any particular draft, draw a horizontal line through that draft. Then, through the point where this cuts the 45° line, draw a vertical line to meet the B and M curves. The scale readings opposite the points thus found give the positions of the centre of buoyancy and the metacentre, whilst the distance between the points is equal to the BM.

For example, in Fig. 79, to find the KB, KM and BM, at a draft of 5·50 metres. Draw the horizontal line, AB, through the 5·50 metre mark and then the vertical line, CBD. C then represents the height of the centre of buoyancy, D that of the metacentre above the keel and CD, the ship's BM. We can see from the scale that KB equals 3·5 metres, KM equals 6·8 metres and BM is 3·3 metres.

CHAPTER 14

BILGING OF COMPARTMENTS

The Effect of Bilging a Compartment.—When a hold or compartment is bilged (i.e., holed, so that it becomes flooded), a number of things can happen.

(*a*) The ship will increase her mean draft in order to compensate for the buoyancy which she has lost, since she must displace her own weight of water in order to float. If an empty hold is bilged, it will cease to displace any water and so the ship must sink until the remaining, intact part of her has made up this loss and displaces a weight of water equal to the weight of the ship. If the hold has cargo in it, such cargo will continue to displace a certain amount of water, so that the bilged compartment only loses a part of its displacement. The amount of displacement then lost, expressed as a percentage of that which would have been lost had the hold been empty, is called the *Permeability* of the hold.

(*b*) If the centre of gravity of the compartment is in the same vertical line as the centre of buoyancy of the ship, the latter will merely sink bodily to a new waterline. If these two points are not in the same vertical line, *B* will shift forward or aft, as the case may be. As the bilging is the cause of loss of buoyancy only and not actual addition of weight to the ship, *G* will not move, so the ship must change her trim in order to bring *B* back into the same vertical line as *G*.

Note the difference between this case and that of weights added, removed, or shifted. In the latter case, *G* moves as well as *B*, so that the relative positions of the weight and the centre of flotation govern whether the ship will change her trim. In this case, where *G* does not shift, the change of trim, if any, is governed by the relative positions of the bilged compartment and the centre of buoyancy.

(*c*) If the compartment is divided longitudinally, the ship may list on account of the lost buoyancy being out of the transverse centre-line.

Permeability.—We have just said that this is the ratio between the space available for water and the total space in the compartment.

For instance, suppose that a compartment has a volume of 5000 cubic metres. This would be the volume available for water if the empty compartment was bilged. If this compartment was filled with cargo, the solid parts of that cargo would take up space which would be otherwise available for water, so that less water would be able to enter the compartment if it was bilged. If the actual solid material in the compartment occupied 3000 cubic metres, then only 2,000 cubic metres of space would be available for water.

$$\text{In this case, the permeability} = \frac{\text{Space available for water}}{\text{Total space}}$$

$$= \frac{2000}{5000} = 0{\cdot}400, \text{ or } 40\%.$$

For certain types of cargo, it is possible to calculate the permeability if the stowage factor and relative density are known:—

Relative density is the ratio between the weight of solid material and the weight of an equal volume of fresh water.

Space occupied by 1 tonne fresh water $= 1 \text{ m}^3$

$$\text{Space occupied by 1 tonne solid cargo} = \frac{1}{\text{Relative density}} \text{m}^3$$

For each tonne of cargo, as stowed, the space available for water = Space occupied by 1 tonne as stowed − Space occupied by 1 tonne solid cargo.

$$= \text{Stowage factor} - (1 \div \text{Relative density})$$

$$\text{Permeability} = \frac{\text{Space available for water}}{\text{Total space}} \text{ (per tonne)}$$

$$\text{Permeability} = \frac{\text{Stowage factor} - (1 \div \text{Relative density})}{\text{Stowage factor}}$$

Example.—Find the permeability of a cargo which has a relative density of 1·80 and stowage factor of 1·20.

$$\text{Permeability} = \frac{1{\cdot}20 - (1 \div 1{\cdot}80)}{1{\cdot}20} = 0{\cdot}537, \text{ or } 53{\cdot}7\%$$

Bilging of an Empty Compartment Amidships.—Fig. 80 represents a vessel in which a compartment amidships has been bilged. The vessel is shown as box shaped, for the sake of simplicity, but the effect will be the same for ship shapes.

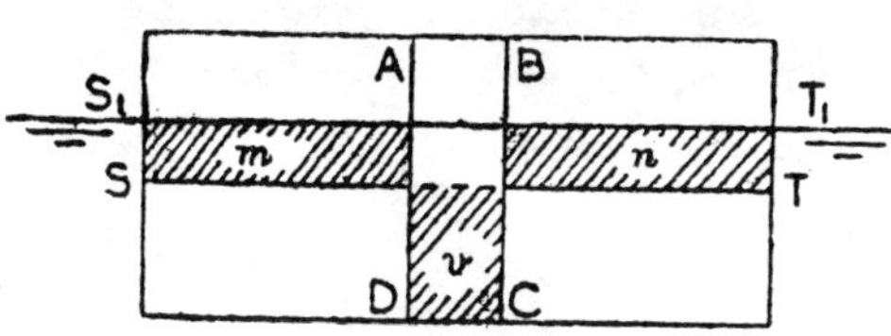

FIG. 80

Suppose that the ship floated originally at the waterline ST and that an empty compartment $ABCD$ is bilged. The buoyancy of this compartment is now lost to the ship. Call this lost buoyancy v. The ship must continue to displace her own weight of water and so must displace the same volume as before. In order to do this, she will sink to the new waterline S_1T_1, so that the total volume of the layers m and n, which have become immersed must be equal to that of the lost buoyancy. This means that:—

$$m + n = v$$

If the total areas of the original and the new waterplanes are the same, the volumes of m and n are together equal to the area of intact waterplane multiplied by the increase of draft (SS_1, or TT_1). So, if A is the total area of waterplane, a the area within the compartment and X the increase of draft, in metres:—

$$m + n = (A - a) X$$

But, $$m + n = v$$

so, $$v = (A - a) X$$

$$X = \frac{v}{A - a}$$

Example.—A box-shaped lighter is 50 metres long, 8 metres wide and floats at a mean draft of 2·00 metres. Find the sinkage if an empty compartment amidships, 20 metres long, is bilged.

Lost buoyancy (v) $= 20 \times 8 \times 2{\cdot}0$ m^3

Original waterplane area (A) $= 50 \times 8$ m^2

Area lost in bilged compartment (a) $= 20 \times 8$ m^2

$$\text{Sinkage} = \frac{v}{A - a} = \frac{20 \times 8 \times 2}{50 \times 8 - 20 \times 8} = 1{\cdot}33 \text{ metres}$$

Bilging of an Amidships Compartment, with Cargo.—If there is cargo in the bilged compartment, it will still continue to displace a certain amount of water and we shall not lose the whole of the buoyancy in the compartment. We shall also gain a little buoyancy by submerging more cargo between the old and new waterplanes.

In this case, if p be the permeability:—

(*a*) The lost buoyancy in the compartment will now become vp cubic metres.

(*b*) In Fig. 80, consider the space between the two waterplanes, ST and S_1T_1 and within the compartment $ABCD$. As the ship sinks to her new waterplane, the cargo in this space will become submerged and will thus add to the buoyancy gained.

The volume of this space $= aX$ cubic metres.

So, space available for water $= apX$ cubic metres.

Buoyancy gained in the compartment = volume − space available for water

$$= aX - apX$$

$$= X(a - ap)$$

And, total buoyancy gained $= m + n + X(a - ap)$

Lost buoyancy = Buoyancy gained

$$vp = m + n + X(a - ap)$$

But (from previous section) $m + n = (A - a)X$

So,
$$vp = (A - a)\,X + X\,(a - ap)$$
$$vp = X\,(A - a + a - ap)$$
$$vp = X\,(A - ap)$$
$$X = \frac{vp}{A - ap}$$

Example.—A box-shaped vessel is 60 metres long, 10 metres wide and floats at a draft of 4·00 metres fore and aft. A compartment amidships is 15 metres long and contains cargo of permeability 40%. Find the new drafts if this compartment is bilged.

$$\text{Volume of compartment } (v) = 15 \times 10 \times 4\ \text{m}^3$$
$$\text{Original waterplane area } (A) = 60 \times 10\ \text{m}^2$$
$$\text{Area in bilged compartment} = 15 \times 10\ \text{m}^2$$
$$\text{Permeability } 40\% = 0{\cdot}40$$
$$\text{Sinkage} = \frac{vp}{A - ap} = \frac{15 \times 10 \times 4 \times 0{\cdot}4}{60 \times 10 - 15 \times 10 \times 0{\cdot}4} = 0{\cdot}44\ \text{m}$$
$$\text{New draft} = 4{\cdot}00 + 0{\cdot}44 = 4{\cdot}44\ \text{metres}$$

Bilging of an Empty Compartment, not Amidships.—In this case, there will be both a bodily sinkage and a change of trim. These may be calculated as follows:—

(*a*) The bodily sinkage is found in the same way as for an amidships compartment.

(*b*) The method of finding change of trim is illustrated by Fig. 81. Here, it can be seen that, since there is a loss of buoyancy at one end of the ship, B will move away from that end to a new position B_1. The forces of buoyancy and gravity then form a couple, forcing the bilged end of the ship downwards. As the ship changes her trim, the centre of buoyancy will move back towards G, until it again comes vertically under G.

It can be seen from the diagram that the couple causing the ship

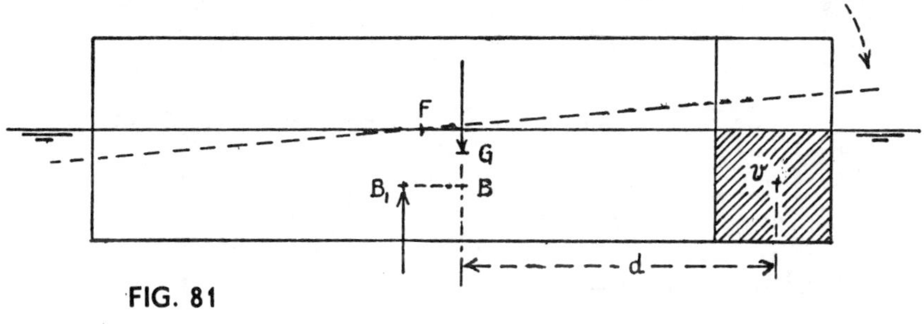

FIG. 81

to change her trim consists of the forces of gravity and buoyancy acting on the lever BB_1.

So, Moment changing trim $= W \times BB_1$

$$\text{Change of trim} = \frac{\text{Moment changing trim}}{\text{M.C.T.1C.}} = \frac{W \times BB_1}{\text{M.C.T.1C.}}$$

(*c*) To find BB_1:—

It has been shewn, in Chapter 5, that $BB_1 = \dfrac{v \times gg_1}{V}$ and that B obeys the same laws as any other centre of gravity.

So if d is the distance of the centre of gravity of the bilged compartment from B, then:—

$$BB_1 = \frac{v \times d}{V}$$

Note.—In the case of an end compartment in a box shaped vessel, BB_1 will be equal to half the length of the bilged compartment. This will only apply to this particular case.

(*d*) If the new drafts are required, we must calculate the change of draft due to trim, as shewn in Chapter 11. Remember, in this case, that the vessel will trim about the *new* centre of flotation, which will be at the centre of gravity of the new, intact waterplane.

Example.—A box-shaped vessel, 100 metres long and 20 metres beam, floats at drafts of 4·00 metres fore and aft. Find the new drafts after an empty end compartment, 10 metres long, is bilged forward.

$$\text{Bodily sinkage} = \frac{v}{A - a} = \frac{10 \times 20 \times 4}{100 \times 20 - 10 \times 20} = 0{\cdot}44 \text{ m}$$

To calculate the M.C.T.1C., we must first find W and BM_L.

$$W = L \times B \times D \times 1{\cdot}025$$
$$= 100 \times 20 \times 4 \times 1{\cdot}025$$
$$= 8200 \text{ tonnes} \qquad \text{(i)}$$

$$BM_L = \frac{I}{V} = \frac{L^2}{12D} = \frac{90^2}{12 \times 4{\cdot}44} = 152 \text{ m} \qquad \text{(ii)}$$

$$\text{M.C.T.1C.} = \frac{W \times BM_L}{100L} = \frac{8200 \times 152}{100 \times 100} = 124 \text{ t/m} \qquad \text{(iii)}$$

$$BB_1 = \tfrac{1}{2} \text{ length of compartment} = 5{\cdot}0 \text{ m} \qquad \text{(iv)}$$

F will be at $\frac{1}{2}$ length of the intact waterplane = 45 m from aft.

$$\text{Change of trim} = \frac{W \times BB_1}{\text{M.C.T.1C.}} = \frac{8200 \times 5}{124} = 331 \text{ cm}$$

Change of draft for trim: Forward $= \frac{55}{100} \times 331 = 182$ cm

Aft $= \frac{45}{100} \times 331 = 149$ cm

Original drafts - -	F.	4·00 m	A.	4·00 m
Bodily sinkage -		0·44 m		0·44 m
		4.44 m		4·44 m
Change for trim - -		+1·82 m		−1·49 m
New drafts - - -	F.	6·26 m	A.	2·95 m

Notes.— (i) The displacement remains unchanged, so we can use, here, the original length and draft.

(ii) I is for the intact part of the waterplane, so if we wish to convert I/V to $L^2/12D$ we must here use the length of the intact waterplane and the new draft.

(iii) Since we do not know GM_L we must use BM_L here.

(iv) This only applies to an empty end compartment in a box-shaped vessel.

The Effect of a Watertight Flat, at or below the Waterline.—If the bilged compartment has a watertight flat at or below the waterline, the sinkage and trim (if any) will be found in the same way as before. From Fig. 82, however, we can see that two modifications will have to be made in the calculations:—

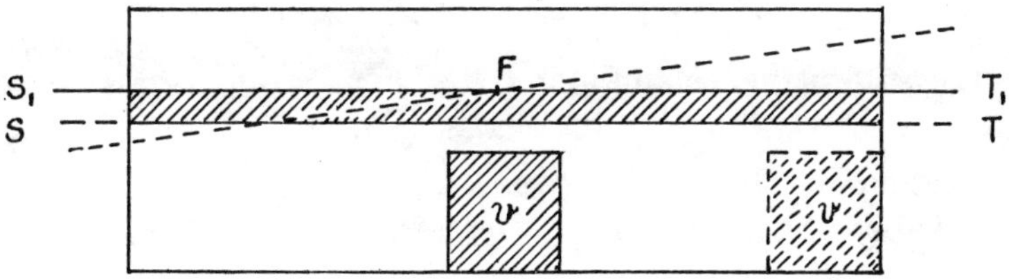

FIG. 82

(*a*) The volume of lost buoyancy (v) will be that below the watertight flat.

(*b*) The whole waterplane will remain intact, so that the buoyancy gained as the vessel sinks will be AX cubic metres (instead of $(A - a)\ X$).

Thus, for an empty compartment: $X = \frac{v}{A}$

Or, if the compartment contains cargo: $X = \frac{Vp}{A}$

(c) BB_1, even for a box shape, must be found by the formula:—

$$BB_1 = \frac{v \times d}{V} \text{ (for an empty compartment)}$$

$$\text{or, } BB_1 = \frac{vp \times d}{V} \text{ (if the compartment contains cargo).}$$

(d) Since the waterplane remains intact, F will not move appreciably when the compartment is bilged; so we can assume that the ship will still trim about the same point as before. In the case of a box shape, this means that the ship will continue to trim about amidships.

Example.—A box-shaped vessel, 120 metres long and 20 metres beam, floats at drafts of 4·00 metres fore and aft. An end compartment, 25 metres long, has a watertight flat 3·00 metres above the keel and contains cargo of permeability 60%. Find the new drafts if this compartment is bilged.

$$\text{Sinkage} = \frac{vp}{A} = \frac{25 \times 20 \times 3 \times 0{\cdot}60}{120 \times 20} = 0{\cdot}38 \text{ m}$$

$$W = 120 \times 20 \times 4 \times 1{\cdot}025 = 9840 \text{ tonnes}$$

$$BM_L = \frac{L^2}{12D} = \frac{120 \times 120}{12 \times 4} = 300 \text{ m}$$

$$BB_1 = \frac{vp \times d}{V} = \frac{25 \times 20 \times 3 \times 0{\cdot}6 \times 47{\cdot}5}{120 \times 20 \times 4} = 4.45 \text{ m}$$

$$\text{M.C.T.1C.} = \frac{W \times BM_L}{100L} = \frac{9840 \times 300}{100 \times 120} = 246 \text{ tonne-metres}$$

$$\text{Change of trim} = \frac{W \times BB_1}{\text{M.C.T.1C.}} = \frac{9840 \times 4{\cdot}45}{246} = 178 \text{ cm}$$

Since the waterplane remains intact, F will remain amidships and the change of draft at either end will be half the change of trim.

Original drafts - -	4·00 m	4·00 m
Bodily sinkage - -	+0·38 m	+0·38 m
	4·38 m	4·38 m
Half change of trim -	+0·89 m	−0·89 m
New drafts - - -	5·27 m	3·49 m

Compartments with Cargo, not Amidships.—Unless such compartments have a watertight flat at or below the waterplane, accurate calculations become very complicated because of the submersion of further cargo as the ship trims. Such calculations are outside the scope of this book.

CHAPTER 15

STABILITY AND THE LOAD LINE RULES

Under the Load Line Rules, ships must conform to stated minimum stability requirements and must also be provided with certain stability information for the use of Masters and Deck Officers.

Stability Requirements.—The ship's stability must be sufficient for the freeboard assigned to her and her light *GM* ascertained by means of an inclining experiment. Also:—

(*a*) The initial *GM* is to be not less than 0·15 metres for ships loaded to ordinary load lines. Ships loaded to timber load lines may, however, have an initial *GM* of not less than 0·05 metres.

(*b*) The maximum righting lever (*GZ*) is to occur at an angle of heel of not less than 30° and must be at least 0·20 metres.

(*c*) The area under the curve of righting levers must not be less than:—

0·055 metre-radians up to 30° of heel.

0·09 metre-radians up to 40° of heel: or up to the angle at which non-weathertight openings become submerged, if this is less than 40°.

0·03 metre-radians between the above.

(*d*) Certain ships may be assigned less freeboard than others, provided that they meet certain additional requirements.

Requirements for Special Types of Ships.—For freeboard purposes, ships are divided into two basic types: Type A, which are ships intended to carry only liquid cargoes in bulk: and Type B, which includes all other ships. If, however, a Type B ship has steel hatch covers, improved protection for the crew, better freeing arrangements and special subdivision against flooding, she may be designated Type B60, or B100.

Ships of Types A, B60 and B100 are allowed to have less freeboard than the basic Type B ship. To qualify for this they must be able to withstand the flooding of one or two compartments, according to their length and type: whilst after such flooding, they must meet the following requirements:—

(*a*) The new waterline must be below any opening through which the ship could become flooded.

(*b*) Any heel due to unsymmetrical flooding (i.e. excess weight of water on one side of the ship), should not be more than 15°. This may be extended to 17° if no part of the deck is then immersed.

(*c*) The ship must have a positive *GM*, when she is upright, of at least 0·05 metre.

(*d*) The range of positive stability must be at least 20°: for example, if the vessel heels to, say, 12° after flooding, her angle of vanishing stability must be not less than 32°.

(*e*) The maximum *GZ* must be at least 0·1 metre.

Information to be Supplied to Ships.—Full details of this may be found in the Load Line Rules. The following is a summary of the requirements:—

(*a*) A plan of the ship to show the capacity and *Kg* of each space: weight and *Kg* of passengers and crew; weight, disposition and *Kg* of any anticipated homogeneous deck cargo.

(*b*) The light displacement and *KG*; also the weight, disposition and *Kg* of permanent ballast, if any.

(*c*) Curves or scales to show displacement, deadweight, *KM*, T.P.C., and M.C.T.1C.

(*d*) A statement of the free surface effect in each tank.

(*e*) Cross curves, stating the assumed *KG*.

(*f*) Statements and diagrams to show displacement, disposition and weights of cargo, etc., drafts, trim information, *KG*, *KM*, *GM*. free surface corrections, and curves of statical stability when the ship is:—

(i) Light.
(ii) In ballast condition.
(iii) Loaded with homogeneous cargo.
(iv) In service loaded conditions.

(*g*) Written instructions concerning any special procedure necessary to maintain adequate stability throughout the voyage.

The Stability Information Booklet.—There is no statutory requirement as to how the specified stability information is to be set out, and this has varied from ship to ship. It would be an advantage to ship's Masters and Officers if a standardised method were used in all ships. To this end, the Department of Trade have produced their own recommended form of Stability Information Booklet.

In order to illustrate the Department's recommendations, the main information for an imaginary ship is set-out in the suggested form in this chapter and in the scales in the back of this book. Some parts have here been abbreviated, or merely described, in order to save space: but it is hoped that this will be sufficient to give the reader a clear idea of the main contents of the booklet, which are as follows:—

(*a*) General particulars of the ship (name, official number, dimensions, tonnage, etc.).

(*b*) Plans of the ship, shewing cargo, tank, store spaces, etc.

(*c*) Special notes regarding the stability and loading of the ship: both in general and as applied to that particular vessel.

(*d*) Hydrostatic particulars for the ship in salt water. (*See* the example given in the back of this book.)

(*e*) Capacities and centres of gravity of cargo spaces, storerooms, crew spaces, etc. (*See* Plate in this chapter.)

(*f*) Capacities, centres of gravity and free surface moments of oil and water tanks. (Also in this chapter.)

(*g*) Notes on the use of free surface moments. (Here described in Chapter 9.)

(*h*) Special information required if the ship is designed to carry containers: including a container stowage plan and a statement indicating the position of the centre of gravity of each container.

(*i*) Cross curves of stability (*KN* curves) and an example shewing their use. (Described here in Chapter 13.)

(*j*) A deadweight scale. (*See* the example given in the back of this book.)

(*k*) Condition sheets, giving a plan and details of weights on board, information on stability on departure or arrival, and a curve of statical stability: all for at least each of the following conditions:—

(i) The light ship.

(ii) Ballast conditions on (*a*) departure and (*b*) arrival.

(iii) The ship loaded to the summer load line with homogeneous cargo on (*a*) departure and (*b*) arrival.

(iv) The ship loaded to the summer load line in at least one service loaded condition on (*a*) departure and (*b*) arrival.

For the above purposes, it is assumed that:—

For each "Arrival Condition", all fuel, fresh water and consumable stores have been reduced to 10% of their original amounts.

In the "departure condition", fuel tanks which are "full" of oil are taken as 98% full.

An abbreviated example of the ship in condition 3 (*a*) is given in this chapter.

CAPACITIES AND CENTRES OF GRAVITY OF CARGO SPACES, STOREROOMS AND CREW AND EFFECTS

(A.B. - ABOVE BASE) (A.P. - AFTER PERPENDICULAR)

DRY CARGO

COMPARTMENT	LOCATION (FRAME NUMBERS)	CAPACITIES		CENTRES OF GRAVITY	
		BALE CUBIC METRES	GRAIN CUBIC METRES	VERTICAL A.B.	LONGITUDINAL FROM A.P.
No.1 TWEEN DECK	146 — 174	656	687	10·43	119·8
No.1 HOLD	146 — 174	1378	1456	5·72	119·3
No.2. TWEEN DECK	118 — 146	947	978	9·88	100·2
No.2 HOLD	118 — 146	2283	2367	5·10	99·8
No.3 TWEEN DECK	70 — 118	1249	1314	9·67	68·8
FOR'D DEEP TANK	105 — 118	1084	1124	4·84	86·1
AFT DEEP TANK	93 — 105	1067	1106	4·76	77·2
No.4 TWEEN DECK	38 — 63	625	650	9·46	34·6

REFRIGERATED CARGO

COMPARTMENT	LOCATION (FRAME NUMBERS)	BALE CAPACITY CUBIC METRES	CENTRES OF GRAVITY	
			VERTICAL A.B.	LONGITUDINAL FROM A.P.
No.4 HOLD	38 — 63	1823	5·22	35·1

STOREROOMS

COMPARTMENT	LOCATION (FRAME NUMBERS)	BALE CAPACITY CUBIC METRES	CENTRES OF GRAVITY	
			VERTICAL A.B.	LONGITUDINAL FROM A.P.
FORECASTLE STORE	174 — 184	34	12·72	135·2
FORE PEAK STORE	174 — 184	51	10·46	134·9
MIDSHIP STORE	63 — 67	50	11·90	48·8
POOP STORE	A — E	68	10·04	−1·8

CREW, STORES AND EFFECTS

COMPARTMENT	DEPARTURE			ARRIVAL		
	TONNES	CENTRES OF GRAVITY VERTICAL A.B	CENTRES OF GRAVITY LONGITUDINAL FROM A.P.	TONNES	CENTRES OF GRAVITY VERTICAL A.B.	CENTRES OF GRAVITY LONGITUDINAL FROM A.P.
STORES	20	9·36	71·4	2	8·88	71·4
CREW & EFFECTS	14	13·87	63·1	14	13·90	63·1

CAPACITIES, CENTRES OF GRAVITY AND FREE SURFACE MOMENTS OF OIL AND WATER TANKS

(A.B. – ABOVE BASE) (A.P. – AFTER PERPENDICULAR)

CARGO-OIL TANKS

COMPARTMENT	LOCATION	CAPACITIES 100% FULL	CAPACITIES 98% FULL		CENTRES OF GRAVITY (METRES)		FREE SURFACE MOMENT
	(FRAME NUMBERS)	CUBIC METRES	CUBIC METRES	TONNES AT S.G. 1·0	VERTICAL A.B. (98%)	LONGITUDINAL FROM A.P.	AT S.G. 1·0
F.P. DEEP TANK	105 – 118	562	551	551	4·84	86·1	639
F.S. " "	105 – 118	562	551	551	4·84	86·1	639
A.P. " "	93 – 105	553	542	542	4·76	77·2	621
A.S. " "	93 – 105	553	542	542	4·76	77·2	621

OIL FUEL TANKS

COMPARTMENT	LOCATION	CAPACITIES 100% FULL	CAPACITIES 98% FULL		CENTRES OF GRAVITY (METRES)		FREE SURFACE MOMENT
	(FRAME NUMBERS)	CUBIC METRES	CUBIC METRES	TONNES AT S.G. 1·0	VERTICAL A.B. (98%)	LONGITUDINAL FROM A.P.	AT S.G. 1·0
No. 2 (P) D.B. TANK	118 – 146	145	142	142	0·76	99·8	138
No. 2 (S) " " "	118 – 146	145	142	142	0·76	99·8	138
No. 3 (P) " " "	93 – 118	228	223	223	0·75	80·3	189
No. 3 (S) " " "	93 – 118	228	223	223	0·75	80·3	189
E.R. (P) " " "	63 – 93	252	247	247	0·74	62·6	171
E.R. (S) " " "	63 --						

ENGINE ROOM TANKS

COMPARTMENT	LOCATION	CAPACITIES 100% FULL	CAPACITIES 98% FULL		CENTRES OF GRAVITY (METRES)		FREE SURFACE MOMENT
	(FRAME NUMBERS)	CUBIC METRES	CUBIC METRES	TONNES AT S.G. 1·0	VERTICAL A.B. (98%)	LONGITUDINAL FROM A.P.	AT S.G. 1·0
E.R. (P) FUEL TANK	89 – 93	114	112	112	3·50	70·5	54
E.R. (S) " "	89 – 93	114	112	112	3·50	70·5	54
LUB. OIL TANK.	63 – 65	24	22	22	5·31	49·0	11

FRESH, FEED AND BALLAST WATER TANKS

COMPARTMENT	LOCATION	CAPACITIES	WATER BALLAST	FRESH WATER	CENTRES OF GRAVITY (METRES)		FREE SURFACE MOMENT
	(FRAME NUMBERS)	CUBIC METRES	TONNES AT S.G. 1·025	TONNES AT S.G. 1·0	VERTICAL A.B. (98%)	LONGITUDINAL FROM A.P.	AT S.G. 1·0
FORE PEAK TANK	174 – 184	113	116	113	4·57	136·2	103
No. 1. D.B. TANK	146 – 174	189	194	189	0·76	119·5	205
F.W. TANK (P)	63 – 70	60	-	60	9·30	50·5	52
F.W. TANK (S)	63 – 70	60	-	60	9·30	50·5	52
AFTER PEAK TANK	0 – 10	163	167	163	7·12	5·1	78

CONDITION SHEET

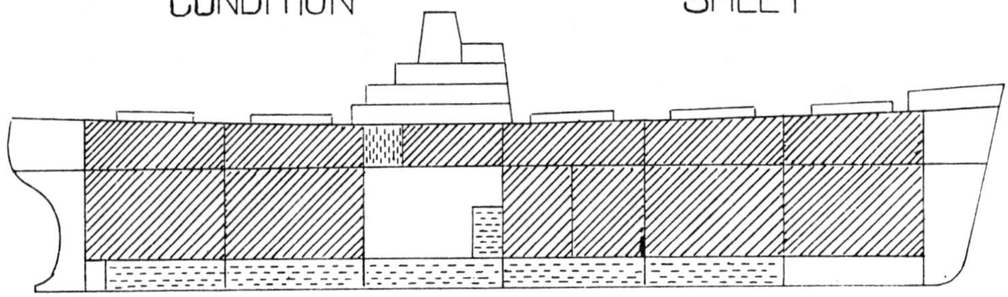

LOADING CODE

DRY CARGO	CARGO OIL
REFRIGERATED CARGO	OIL FUEL
FRESH WATER	WATER BALLAST

CONDITION No. 3A			HOMOGENEOUS LOADING CONDITION — DEPARTURE CARGO OF STOWAGE FACTOR 1·507 M³/T.		
ITEM OF DEADWEIGHT	WEIGHT	V.C.G.	VERTL MOMENT	L.C.G.	LONGL MOMENT
No. 1. T.D.	436	10·43	4547	119·8	52233
No. 1. L.H.	914	5·72	5228	119·3	109040
No 2. T.D.	628	9·88	5205	100·2	62926
No. 2. L.H.	1515	5·60	8484	99·8	151197
No. 3. T.D.	829	9·79	7867	68·8	57035
F. D TANK.	719	4·84	3480	86·1	61906
No. 2. D.B.	256	0·76	195	99·8	25549
No. 3. D.B.	402	0·75	302	80·3	32281
E.R. TANK.*	202	3·50	707	70·5	14241
E.R.D.B. *	444	0·74	329	62·6	27794
F.W. TANKS *	118	9·20	1086	50·5	5959
CREW STORES	20	9·40	188	71·4	1428
CREW.	14	13·90	195	63·1	883
DEADWEIGHT.	10720	6·395	68557	70·200	752543
LIGHT SHIP	4559	6·623	30194	72·687	331350
DISPLACEMENT	15279	6·463	98751	70·942	1083923

SAILING STATE	
DRAFT AT L.C.F.	8·094
DRAFT AFT	8·319
DRAFT FORWARD	7·853
MEAN DRAFT	8·086
TRIM BY STERN	0·466
K.M.	8·025
K.G.	6·463
G.M. (SOLID)	1·562
FREE SURFACE CORRN	0·034
G.M. (FLUID)	1·528

* TANK WITH FREE SURFACE EFFECT.

The Use of Maximum Deadweight Moments.—This is a simplified method of making sure that a ship's *GM* is large enough to enable her to comply with, at least, the minimum requirements set out in the Load Line Rules. It is intended for use in small ships, for which it is particularly suitable.

The method is, basically, an adaptation of that described in Chapter 5 for finding a ship's *KG* by taking moments about the base line. In that chapter, it has been shewn that:—

(*a*) Deadweight Moment is the sum of the moments about the base line of all weights on board the ship (i.e. cargo, fuel stores, etc.; which are the components of deadweight).

(*b*) The total moment about the base line equals the moment of the light ship plus the deadweight moment.

(*c*)
$$KG = \frac{\text{Total moment}}{\text{Displacement}}$$
$$= \frac{\text{Moment of light ship} + \text{Deadweight moment}}{W}$$

Now, the moment of the light ship is equal to the light displacement multiplied by the light *KG*, which is a fixed quantity for a given ship. The displacement depends on the ship's draft and will be known. So we can use the above formula to calculate that deadweight moment which will give the ship a desired *KG*.

$$\text{But,} \quad GM = KM - KG$$

KM depends on the ship's draft and will also be known

$$\text{So,} \quad GM = KM - \frac{\text{Moment of light ship} + \text{Deadweight moment}}{W}$$

This means that we can calculate that deadweight moment which will give us a desired *GM* for a ship at given draft.

In practice, the Naval Architects first find the least *GM* which, at a given draft, will satisfy the requirements of the Load Line Rules. They then calculate the deadweight moment which will give this *GM*: this will be the maximum allowable deadweight moment for the chosen draft (maximum because the greater the moment, the greater the *KG* and the less the *GM*).

Example.—A small vessel has a light displacement of 737 tonnes and a light *KG* of 3·300 m. At 3·00 m draft, her displacement is 1259 tonnes and her *KM* is 3·845 m. It is calculated that, at this draft, her *GM* must be at least 0·238 m in order to comply with the Load Line Rules. Find her maximum allowable deadweight moment at 3·00 m draft.

Required minimum	*GM*	0·238m
	KM	3·845m
Max. allowable	*KG*	3·607 m

$$\text{Maximum total moment} = W \times KG = 1259 \times 3{\cdot}607 = 4541 \text{ t/m}$$
$$\text{Moment of light ship} = 737 \times 3{\cdot}300 = 2432 \text{ t/m}$$

$$\text{Maximum permissible deadweight moment} = 2109 \text{ t/m}$$

The above is repeated for a series of drafts between the light and load waterlines and, from this, a scale or graph is drawn up to shew the maximum permissible deadweight moment for each draft. The seaman is given a copy of this scale and/or graph: also a form on which is shewn a profile of the ship and the heights of the centres of gravity of the various compartments. He enters on this form, the amount and Kg of each item on board and multiplies them together to find its deadweight moment. The sum of these moments will be the actual deadweight moment of the ship. The seaman also extracts from the scale or graph, the maximum permissible deadweight moment for his ship's draft or displacement. As long as the actual deadweight moment is less than the maximum permissible moment, the ship will have a sufficient GM.

An example of the above is shewn in the following two diagrams. The first of these shews the maximum permissible deadweight moments for an imaginary small ship. The second shews a completed form for the same ship, when loaded: indicating that at a displacement of 1861 tonnes, the ship has an actual deadweight moment of 3702 tonne-metres. The maximum permissible deadweight moment for the ship's displacement is then extracted from the first diagram (in this case it is 4141 t/m): this is entered at the bottom of the second form. This then shews that, in this case, the ship has sufficient GM, since the actual moment is less than the maximum permissible moment.

"Simplified Stability Information".—This may be provided as an addition to the basic data and sample loading conditions required by the Rules. This information may be presented in one of three ways, provided that it is accompanied by clear guidance notes for its use:—

(*a*) A maximum deadweight diagram or table.
(*b*) A diagram or table shewing maximum permissible KGs.
(*c*) A diagram or table shewing minimum permissible GMs.

The method of setting out the diagrams or tables for KGs or GMs would be basically similar to those shewn here for deadweight moments.

SIMPLIFIED STABILITY INFORMATION

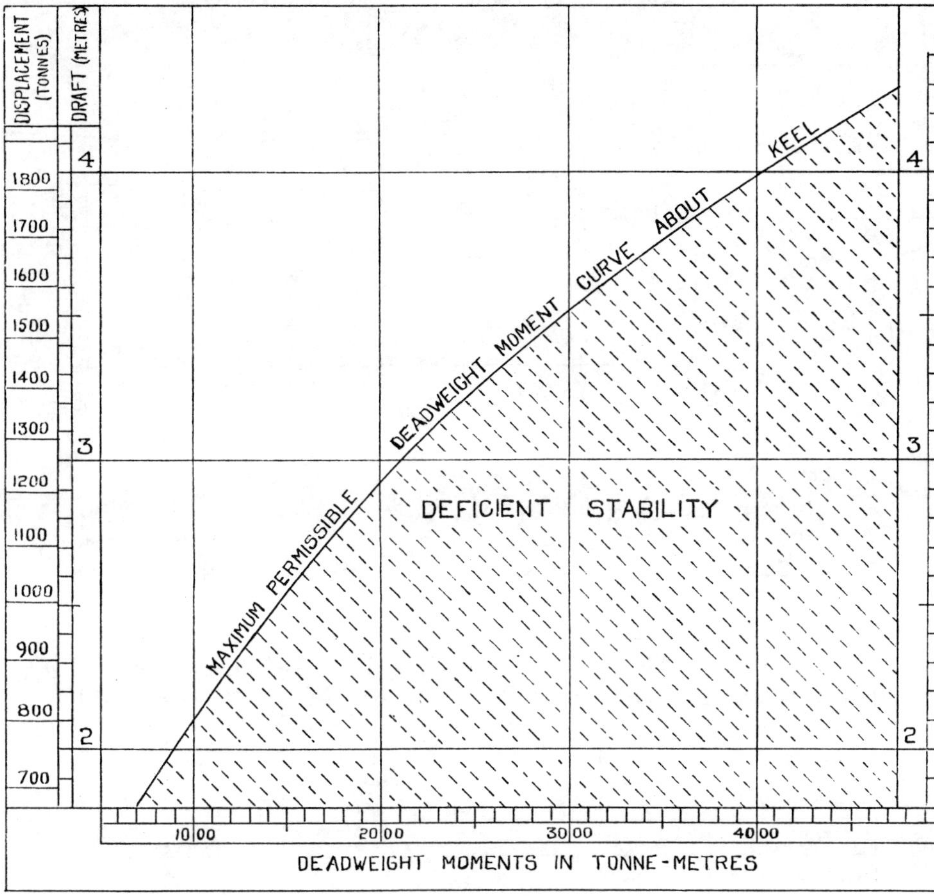

MAXIMUM PERMISSIBLE DEADWEIGHT MOMENT (TONNE - METRES)	2109	2034	1961	1891	1823	1757	1693	1630	1568	1508	1449	1391	1334	1277	1221	1165	1110	1055	1001	948	896	845	796	749
DISPLACEMENT (TONNES)	1259	1232	1205	1178	1151	1124	1098	1072	1046	1020	994	969	944	919	894	870	846	823	800	778	756	735	714	693
DRAFT (METRES)	3·00	2·95	2·90	2·85	2·80	2·75	2·70	2·65	2·60	2·55	2·50	2·45	2·40	2·35	2·30	2·25	2·20	2·15	2·10	2·05	2·00	1·95	1·90	1·85

MAXIMUM PERMISSIBLE DEADWEIGHT MOMENT (TONNE - METRES)	4524	4402	4281	4161	4042	3293	3806	3691	3577	3469	3363	3258	3153	3054	2958	2864	2773	2684	2597	2512	2429	2347	2266	2186
DISPLACEMENT (TONNES)	1959	1928	1897	1866	1836	1805	1775	1745	1715	1685	1656	1627	1598	1569	1544	1511	1483	1454	1426	1398	1370	1342	1314	1286
DRAFT (METRES)	4·20	4·15	4·10	4·05	4·00	3·95	3·90	3·85	3·80	3·75	3·70	3·65	3·60	3·55	3·50	3·45	3·40	3·35	3·30	3·25	3·20	3·15	3·10	3·05

SIMPLIFIED STABILITY INFORMATION

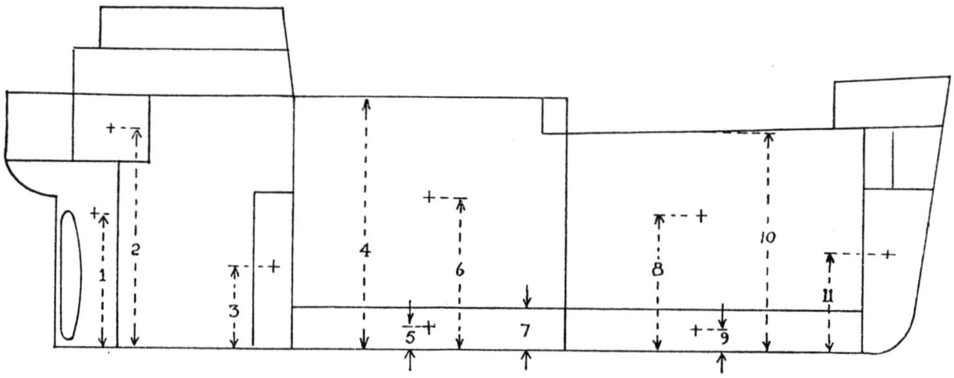

HEIGHTS - 'h'

1 = 2·67m.	2 = 6·03m.	3 = 2·24m.	4 = 6·00m.
5 = 0·52m.	6 = 3·58m.	7 = 1·00m.	8 = 3·12m.
9 = 0·55m.	10 = 5·02m.	11 = 2·36m.	

ITEMS OF DEADWEIGHT	WEIGHT (ω)	HEIGHT ABOVE KEEL TO CENTRE OF WEIGHT-'h'	DEADWEIGHT MOMENT - (ω×h)
CARGO IN FORE HOLD	437	3·12	1363
CARGO IN AFT HOLD	597	3·58	2137
CARGO ON DECK			
OIL FUEL	90	2·24	202
W.B. IN FOREPEAK			
W.B IN AFTER PEAK			
LIQUID IN F. C.D.B.			
LIQUID IN A.C.D.B.			
AFT. STORE			
TOTAL DEADWEIGHT	1124		3702
LIGHT SHIP	737	3·300	2432
DISPLACEMENT	1861		6134

N.B. — AT THIS DISPLACEMENT, THE TOTAL DEADWEIGHT MOMENT MUST NOT EXCEED THE VALUE OF 4141 T/M, AS SHOWN BY THE DEADWEIGHT MOMENT SCALE.

CHAPTER 16

MISCELLANEOUS MATTERS

Drydocking.—When a ship is drydocked, her support has to be transferred from the water to the keel blocks and shores. She may be considered safe whilst she is waterborne, or once the shores have been set up, but there is a danger that she may become unstable during the intervening period, which is often termed the "critical period".

Whilst the dock is being pumped out, the ship at first sinks bodily as the water-level falls, but as soon as she touches the keel blocks she stops sinking and the water falls around her. She thus loses displacement so that weight, equal to the amount of the lost displacement, is transferred to the blocks. As far as the ship's stability is concerned, this weight is equivalent to a force acting vertically upwards at the keel and it will decrease the metacentric height. The latter must, sooner or later, become negative and if this were to happen before the shores were properly set up, the ship might capsize in the dock. It is thus of the utmost importance to keep full control of the ship during the critical period and to get the shores set up as soon as possible. To assist in this, it is usual to have the ship trimmed a little by the stern when she enters the dock, so that the heel of the stern post is the first part to touch the blocks.

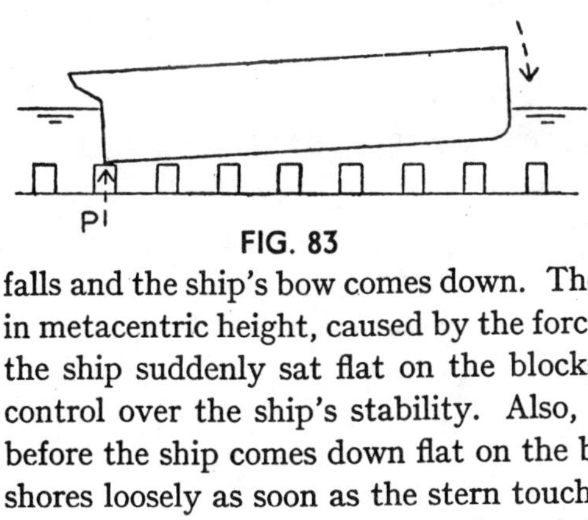

FIG. 83

Fig. 83 illustrates what happens in the above case. As soon as the ship's stern touches the blocks the upward force, P, comes into existence. This force is small at first, but gradually increases as the water-level falls and the ship's bow comes down. The advantage of this is that the decrease in metacentric height, caused by the force P, is more gradual than it would be if the ship suddenly sat flat on the blocks fore and aft, so that we have more control over the ship's stability. Also, although the shores cannot be set up before the ship comes down flat on the blocks, we can start to put-in the after shores loosely as soon as the stern touches. By the time that the ship is right down on the blocks a large number of shores are already in place, so that the remainder can be put in and all set up with the minimum of delay. This decreases the risk of the ship falling over in the dock.

It is important to have the ship upright when she enters a drydock. If she were not, this could be due to one of two causes: a negative metacentric height, or the weights on board not being symmetrical about the centre-line. In the first case, the ship would be certain to fall over as soon as her keel touched the blocks. In the second, she might fall over at some time during the critical period on account of the excess of weight on one side.

Before the ship is floated again, it is very important to check any weights which may have been shifted whilst she is in the dock; otherwise we may have a similar effect to the above whilst the dock is being filled. In this respect, do not forget to make sure that boilers have not been filled or emptied, or to check-up on any weights shifted in the engine-room.

The procedure of dry docking is, briefly, as follows. As soon as the ship enters the dock she usually comes under the control of the foreman carpenter or shipwright, who manoeuvres her into the position he requires. The dock gates are then closed and pumping-out commences. When the ship's stern is nearly on the blocks, pumping is stopped whilst the ship is aligned so that her centre-line is exactly over them. Pumping is then resumed slowly until the stern touches the blocks, when the after shores are put-in loosely. As the ship settles down, more shores are put-in, working from aft forward, and as soon as the keel comes flat on the blocks any remaining shores are put in place and all are set-up as quickly as possible. The heads of shores should always be placed on frames and not between them, in order to eliminate the risk of denting the ship's plating. Once the shores have been set-up, pumping is continued quickly until the dock is dry.

The following formula will give the ship's metacentric height at any time during the process of drydocking:—

Where P = the force acting upwards through the keel.
KM = height of the metacentre on entering the dock.
W = ship's displacement on entering dock.

$$\text{New } GM = \text{Old } GM - \frac{P \times KM}{W}$$

The force P is the difference between the displacement of the ship on entering the dock and her displacement at the time for which we wish to calculate her GM. After the ship has come flat on the blocks, this calculation is quite simple, since the two displacements will be those for the respective mean drafts: that is:—

P = displacement at original draft − displacement at new draft.

If the displacement has decreased much, it may cause M to rise appreciably. If so, we should add a correction of $MM_1\left(1 - \frac{P}{W}\right)$ to the new GM.

It is more difficult to find P during the critical period, after the ship's stern has touched the blocks, but before she comes flat on them. The most dangerous part of this period, and hence the one with which we are most concerned, occurs at the instant before the ship takes the blocks fore and aft. For this instant, P can be found approximately by the following formula:—

Where t = the trim, in centimetres, on entering the dock
l = the distance between the after block and the centre of flotation.

$$P = \frac{\text{M.C.T.1C.} \times t}{l}$$

Example.—A ship entered a drydock with drafts of 3·00 metres forward and 3·50 metres aft. At this draft, her displacement was 4650 tonnes; KG, 7·4 metres; KM, 9·2 metres; M.C.T.1C., 148 tonne-metres; whilst the centre of flotation was 72 metres from aft. Find:—

(1) The GM at the instant before the ship came flat on the blocks.

(2) The GM when the water had fallen so that the draft read 2·65 metres, her displacement was 3650 tonnes, and KM was 10·0 m.

(1) Trim entering dock = 50 cm by stern

GM entering dock = $KM - KG = 9{\cdot}2 - 7{\cdot}4 = 1{\cdot}8$ m

$$P = \frac{\text{M.C.T.1C.} \times t}{l} = \frac{148 \times 50}{72} = 103 \text{ tonnes}$$

$$\text{New } GM = \text{Old } GM - \frac{P \times KM}{W}$$

$$= 1{\cdot}80 - \frac{103 \times 9{\cdot}2}{4650} = 1{\cdot}60 \text{ metres}$$

(2)

$$P = \text{Old displacement} - \text{New displacement}$$
$$= 4650 - 3650 = 1000 \text{ tonnes}$$

$$\text{New } GM = \text{Old } GM - \frac{P \times KM}{W} + MM_1\left(1 - \frac{P}{W}\right)$$

$$= 1{\cdot}80 - \frac{1000 \times 9{\cdot}2}{4650} + 0{\cdot}8\left(1 - \frac{1000}{4650}\right)$$

$$= 1{\cdot}80 - 1{\cdot}97 + 0{\cdot}62 = 0{\cdot}45 \text{ metres}$$

Grounding.—When a ship runs ashore, her metacentric height will decrease or become negative as in drydocking, but the exact effect of this on her stability is almost unpredictable. It will vary according to the nature of the ground, how the ship is placed on the bottom, what damage she has sustained and the nature and state of the tides. In practice, we can only attempt to get the ship afloat again as soon as possible, if it appears safe to do so.

The Effect of Density on Stability.—When there is a change in the density of the water in which a ship floats, she will change her draft. An increase of density will generally have the following effects on her stability:—

(*a*) When the ship is near her light draft, the centre of buoyancy will be lowered and the metacentre will probably rise. Thus there will be an increase of *BM* and probably of *GM* also.

(*b*) When the ship is near her load draft, the centre of buoyancy will fall and the metacentre will probably do likewise. There will usually be little change in *BM*, but *GM* may decrease.

(*c*) The trim may change, usually so as to cause an increase of the draft aft and a decrease forward.

It should be remembered that all the above effects are comparatively small, so that they can safely be ignored in practice.

The Effect of Density on the Draft of Ships.—In Chapter 4, we considered this effect in box shapes and also with regard to fresh water allowance Now let us consider it in relation to ship shapes.

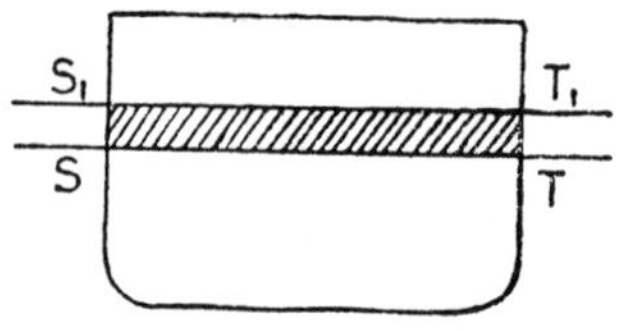

FIG. 84

In Fig. 84, let the ship be first floating at the waterline *ST*, in water of density δ. Then let her pass into water of a lesser density, δ_1 and sink to the new water-line S_1T_1. Let X be the increase of draft, in centimetres — that is, the distance between the two water-lines. Let W be the ship's displacement: since the ship must always displace her own weight of water, this will remain the same for both drafts.

$$\text{Volume} = \frac{\text{Weight}}{\text{Density}}$$

$$\text{So, volume below the water-line } ST = \frac{W}{\delta}$$

$$\text{And volume below the water-line } S_1T_1 = \frac{W}{\delta_1}$$

The difference between these two volumes must be equal to the volume of the layer SS_1T_1T, so:—

$$\text{Volume of the layer } SS_1T_1T = \frac{W}{\delta_1} - \frac{W}{\delta}$$

$$= \frac{W\delta - W\delta_1}{\delta\delta_1}$$

$$= \frac{W}{\delta\delta_1}(\delta - \delta_1) \quad . \quad . \quad . \quad (1)$$

The weight of a layer such as SS_1T_1T must be equal to its depth in centimetres multiplied by the tonnes per centimetre immersion.

$$\text{Weight of the layer } SS_1T_1T = X \times \text{T.P.C.}$$

$$\text{Volume of the layer } SS_1T_1T = \frac{X \times \text{T.P.C.}}{\delta} \quad . \quad . \quad . \quad (2)$$

If we now combine formulae (1) and (2) above, we can see that:—

$$\frac{X \times \text{T.P.C.}}{\delta} = \frac{W}{\delta\delta_1}(\delta - \delta_1)$$

$$X \times \text{T.P.C.} = \frac{W}{\delta_1}(\delta - \delta_1)$$

$$X = \frac{W(\delta - \delta_1)}{\delta_1 \times \text{T.P.C.}}$$

Derivation of the Fresh Water Allowance.—The F.W.A. is the amount, in millimetres, by which a ship would change her draft on passing from salt to fresh water, or vice-versa, when floating at, or very near, her Summer Loadline.

Let the displacement and T.P.C. at summer draft be Δ and T respectively. A change of density from salt to fresh water will give δ as 1025 kg/m³ and δ_1 as 1000 kg/m³. Substituting these in the above formula gives:—

$$\text{Change of draft } (X) = \frac{\Delta(1025 - 1000)}{1000T} \text{ cms.}$$

But F.W.A. is expressed in millimetres, so we must multiply the result by 10, to convert it from cms.

So,

$$\text{F.W.A.} = \frac{\Delta(1025 - 1000)}{1000T} \times 10 \text{ mm.}$$

$$\text{F.W.A.} = \frac{\Delta}{4T} \text{ mm.}$$

Reserve Buoyancy.—In the case of a ship, this is the volume of the hull between the water-line and the freeboard deck. It amounts, approximately to the difference between the actual displacement and that which the ship would have if she were submerged to her freeboard deck.

We can calculate the reserve buoyancy for any floating body by finding the difference between the total watertight volume of the body and the volume of water which it displaces.

Continuous Watertight Longitudinal Bulkheads.—These give great longitudinal strength to a ship and also reduce free surface effect when liquids are carried in bulk. They have one serious disadvantage, however, in that if the ship is holed on one side and the bulkhead remains intact, the compartment could become flooded on one side only. This would give the ship a list, which may be dangerous if the compartment is large.

In ordinary cargo ships, having large holds, there would be considerable risk of the ship capsizing in the above circumstances. There is normally no free surface effect to be reduced in the holds and the bulkheads have the additional disadvantage that they interfere with the handling of cargo. Consequently continuous longitudinal bulkheads are not fitted in ordinary cargo ships, since the disadvantages outweigh the gain in longitudinal strength.

In the case of oil-tankers, carrying bulk liquid cargoes, some form of longitudinal subdivision is necessary to minimise free surface effect. Interference with the stowage of cargo does not have to be considered and great longitudinal strength is required. In such ships, the advantages of continuous longitudinal bulkheads are obvious and one or two are always fitted. The danger of the vessel's capsizing in the event of her being bilged is overcome by restricting the length of her tanks. Also, in the event of a tank on one side becoming flooded, the corresponding tank on the other side could be filled quickly to counterbalance this.

Non-Continuous Longitudinal Bulkheads.—These are often fitted in ordinary ships, as they have a number of structural advantages. Since they are not continuous throughout any hold, they do not affect the ship's stability.

Bulkhead Subdivision and Sheer.—The subdivision of a ship into compartments by means of transverse bulkheads is a great factor in determining her safety if she is holed. It is not generally realised by seamen that sheer also plays an important part in this if the ship is holed forward or abaft the centre of the flotation.

In 1912, a committee was set up to investigate the spacing of bulkheads and the suggestions which were made in their report are now compulsory for passenger ships. It was not possible to apply them to cargo ships also and the bulkheads in the latter are usually more widely spaced than would be allowed in passenger vessels. The committee introduced the "Margin Line" and the "Curve of Floodable Lengths".

The Margin Line is an imaginary line, 75 millimetres below the bulkhead deck. It is assumed that a ship which was sunk to this line would still be navigable in fine weather.

The Curve of Floodable Lengths is a graph from which can be found the floodable lengths for any part of the ship, i.e. that length of the ship which, if flooded, would cause her to sink to her margin line. When this is calculated, allowance is made for an assumed average permeability in each of the various compartments. The length allowed for any compartment is found by multiplying the floodable length by a factor which depends on the length of the ship and on a number of other things.

Sheer is the upward rise of the ship's deck from amidships towards the bow and stern. If a compartment becomes bilged at one end of the ship so that she changes her trim, the sheer helps to prevent the margin line from becoming submerged at that end. It thus increases the floodable lengths forward and aft.

Pressure on Bulkheads.—When a compartment is flooded, the water pressure on the end bulkheads is greatest at the bottom and decreases to nothing at the water-level. The greatest support is needed at a point somewhere between the bottom and the surface level, such that the total pressure above it is equal to the total pressure below it. This point is known as the "Centre of Pressure" and its position is as follows:—

(*a*) For rectangular bulkheads, at the two-thirds depth of the bulkhead below the water-level.

(*b*) For triangular bulkheads (apex downwards), at the half-depth of the bulkhead below the water-level.

(*c*) In the case of ship shapes, the majority of the bulkheads are nearly rectangular and need shoring most strongly at about the two-thirds depth. Some of the bulkheads are, however, of an intermediate shape and in this case the centre of pressure falls between the half-depth and the two-thirds depth.

Three formulae may be useful here:—Where h is the depth below the surface of any point on a bulkhead; H, the depth below the surface of the centre of gravity of the underwater part of the bulkhead; A, the underwater area of the bulkhead; W, the weight of the water in kg per m³:—

$$\text{Pressure at any given point} = Wh \quad (\text{kg per m}^2)$$
$$\text{Average pressure on the bulkhead} = Wh \quad (\text{kg})$$
$$\text{Total pressure on the bulkhead} = WHA \quad (\text{kg})$$

Note that H, which is sometimes called the "Centroid", must not be confused with the centre of pressure. For instance, in a rectangular bulkhead, the centre of pressure is at the two-thirds depth, whilst the centroid is at the half depth below the water-level.

CHAPTER 17

ROLLING

The Formation of Waves.—Waves are produced by friction between the wind and the sea surface. The wind blows, to a greater or less degree, in gusts and also appears to blow somewhat obliquely down on to the sea surface. The effect of this is to cause slight depressions in that surface in some places, with corresponding elevations elsewhere so that "ripples" are formed. The wind will now act directly on these ripples, and if it blows long enough and strongly enough it will turn them into waves. It appears that, within certain limits, the size of the waves will depend largely on the force of the wind and on the distance from the point at which the waves originated.

The Trochoidal Theory.—This theory is generally used to explain the construction of waves and also certain phenomena connected with them. It is rather too complicated for us to consider fully here, so we will merely extract two points from it.

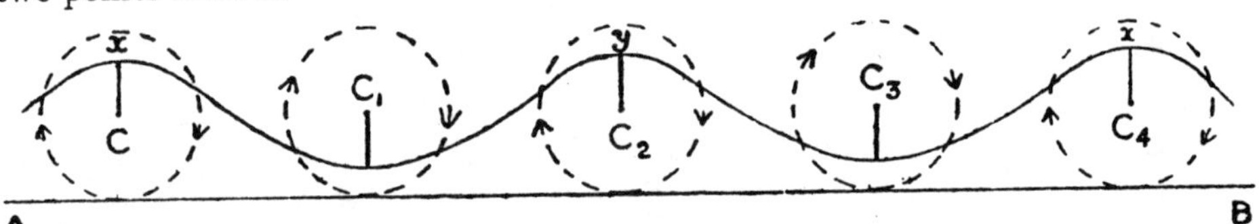

FIG. 85

(*a*) The shapes of waves are approximately the same as a "trochoid", which is the curve traced out by a point inside a circle, rolling along a straight line. This is shown in Fig. 85. Suppose that a wheel, with centre C, be rolled along the level surface *AB*, from left to right. A point, *x*, on the wheel would trace out the trochoid *xyx*, which has approximately the same shape as the surface of the waves—notice that the crests are sharper than the troughs.

(*b*) The water in a wave is not considered to have any appreciable horizontal motion; that is, it does not travel along with the wave in a horizontal direction. Each particle of water moves in a circular orbit, forward at the crest, backward in the trough, upwards in front of the wave and downward behind it. This produces a progressive "heaping-up" of the water, which causes the wave-outline to travel along, although the water itself is not doing so. For our purposes, we can consider waves as comparatively shallow, vertical movements of water.

The True Period of Waves.—This is the interval between the passage of any two consecutive wave crests at a stationary point. If a ship were stopped and had no movement, the period of the waves would be the interval between the time she was on one wave-crest and the time she was on the next. In theory, it is often assumed that a series of waves all have the same period; in practice, this very rarely occurs and successive groups of waves often have slightly different periods. The period usually increases with the size of waves, but rarely exceeds ten or twelve seconds.

The Apparent Period of Waves.—When a ship is moving through the water, the period of the waves may appear to those on board to be greater or less than the true period. A ship which is steaming head-on into a sea will be moving to meet each successive wave, which will thus reach her more quickly and will appear to have a shorter period than it actually has. A ship which has the sea aft, on the other hand, is moving away from the waves, so that these will take longer to catch up with her and will appear to have a longer period. When the sea is exactly abeam of the ship, her motion will have no effect and the apparent period will be the same as the true period. The apparent period will thus depend on the ship's speed through the water and on her course relative to the direction of the waves.

The apparent period is important, because it is the one which is actually felt by the ship and which thus affects her rolling.

The Period of a Ship.—This is the time taken by a ship to roll from one side to the other and back again. When the period is exactly the same for every roll, the rolling is termed "isochronous". It is often assumed that isochronous rolling occurs in every ship for any angle of roll, but this is not correct. We may accept the following general rules:—

(*a*) Different ships have different periods of roll.

(*b*) The same ship will have a different period for different conditions of loading.

(*c*) The same ship will have a longer period when she is tender than when she is stiff.

(*d*) "Winging out" weights will increase the period, all other things being equal.

(*e*) Rolling is isochronous for small angles of roll, up to about ten degrees, but the period increases slightly for larger angles.

Synchronism.—This is said to occur when the ship's period of roll is the same as the apparent period of the waves. When it occurs, the waves give the ship a "push" each time she rolls, in the direction in which she is rolling, causing her to roll more and more heavily. In theory, it would continue until she capsized, but this does not happen in practice because of certain resistances, which we shall consider later.

Unresisted Rolling.—Let us assume for a moment that there are no forces in existence to damp a ship's roll and that she merely rolls under the influence of the waves and of her own period. Let us also assume that the period of the waves is the same throughout and that the ship's period is isochronous for all angles of roll.

If the ship's period is much less than that of the waves, she will always take up a position at right angles to the wave-slope and will not roll to either greater or smaller angles than this. In other words, she will behave in exactly the same way as a raft would do.

If synchronism occurs, the ship will roll through increasingly greater angles until she capsizes.

When the ship's period is much greater than that of the waves, she will roll easily and never to large angles. The waves would set her rolling, but would soon become out of time with the roll and would thus cause her to steady-down again.

In practice, we find that the ship's period increases with the angle of roll and also that we rarely meet a long series of waves of exactly the same period. So, even if the ship herself offered no resistance to rolling, synchronism would seldom exist for long. A normal ship might roll heavily, but she would be unlikely to capsize.

Resistances to Rolling.—We all know that it is possible to set a boat rolling by leaning from side to side in time with her period. If we first start her rolling and then sit upright and without moving, the roll will gradually decrease and will finally die out altogether.

A ship can be set rolling in still water, in a similar way, by shifting weights or bodies of men from side to side. Similarly also, if we stop moving them, the ship's roll will die out. For this to happen, the boat or ship must be setting up resistance to its own roll; otherwise it would merely continue to roll to the same angle and with the same period. These resistances are usually considered to be as follows:—

(*a*) A rolling ship creates waves and these require a considerable amount of work to produce them. This work is provided by the ship and would otherwise have formed part of the forces producing the rolling. Instead, it now passes away with the waves and is lost. This wave-formation is one of the chief resistances to be considered.

(*b*) Friction between the water and the hull of the ship sets up a slight resistance. The effect is very slight in a ship having a much rounded bottom. In a ship which has a full bilge, or particularly bilge keels, it is much greater but is never very considerable.

(*c*) There is a certain amount of resistance between the hull and the air, but its effect is negligible.

(*d*) Properly fitted bilge keels have a great damping effect on rolling. Their exact effects are complicated and will be considered more fully in the next section.

It can be seen that since the above resistances are capable of eliminating the ship's roll in still water, they will also resist the forces causing the ship to roll in a seaway. They will not eliminate rolling, but they can damp it considerably.

The Effects of Bilge Keels.—These may be said to have three main effects in resisting rolling, namely:—

(*a*) They offer a certain amount of direct resistance to the water, but this effect is comparatively weak.

(*b*) They cause the ship's period of roll to increase slightly.

(*c*) They set up eddy currents and pressure under water.

The latter effect is by no means simple and may be considered to produce a number of subsidiary effects:—

1. The wave formation due to rolling, described in subdivision (*a*) of the last section, is considerably increased.

2. The water pressure on the hull is increased on that side of each bilge keel towards which the ship is rolling. This pressure acts at right angles to the hull and its direction is such that it forms a resistance to the roll of the ship.

3. Water is unable to run around the hull in an uninterrupted streamline. This will also reduce rolling, since any upsetting of streamline effects will cause eddy currents and resistance to motion.

Bilge keels have a greater effect when the ship is moving than when she is stationary, and the greater the speed the greater the effect. This is often considered to be due to the fact that as the ship moves ahead she is passing out of the water which has been disturbed by her rolling. A certain amount of the bilge keel is thus working in undisturbed water and the effect of this part is increased accordingly. The faster the ship moves, the more of the keel is in such water and the greater the anti-rolling effect. There may also be some "planing" effect.

Cures for Heavy Rolling.—When, for any reason, a ship is found to be rolling heavily, the proper cure is to alter course and/or speed. This will alter the apparent period of the waves and destroy synchronism, which is nearly always the cause of such rolling.

It is usually unwise, and sometimes dangerous, to attempt a cure by working water ballast in these circumstances.

CHAPTER 18

SUMMARY

ABBREVIATIONS

a	-	Waterplane area in bilged compartment.
A	-	Area of waterplane.
A/P	-	The after perpendicular.
b	-	Breadth of the ship.
B	-	Centre of buoyancy.
d	-	Distance (of shift of weights, etc.)
D	-	Depth, or draft.
F	-	Centre of flotation.
Fa	-	Fresh water allowance.
g	-	Centre of gravity of weight or wedge.
G	-	Centre of gravity of ship.
GM	-	Transverse metacentric height.
GM_L	-	Longitudinal metacentric height.
GZ	-	Righting lever.
h	-	Common interval (Simpson's rules, etc.).
hh_1	-	Horizontal displacement of the centres of gravity of wedges.
i	-	Moment of inertia of free surface.
I	-	Transverse moment of inertia of a waterplane.
I_L	-	Longitudinal moment of inertia of a waterplane.
K	-	Denotes the keel.
KB	-	Height of the centre of buoyancy above the keel.
KG	-	Height of the centre of gravity above the keel.
Kg	-	Height of the centre of gravity of a weight above the base line.
KM	-	Height of the transverse metacentre above the keel.
KM_L	-	Height of the longitudinal metacentre above the keel.
KN	-	The righting lever, assuming G to be at the keel.
l	-	Length.
L	-	Length.
M	-	Transverse metacentre.
M_L	-	Longitudinal metacentre.
M.C.T.1C.	-	Moment to change trim one centimetre

p - Permeability of a compartment.

t - Trim.

T - T.P.C.

T.P.C. - Tons per centimetre immersion.

v - Volume of immersed or emerged wedge, or volume of buoyancy

V - Volume of displacement of a ship.

w - Weight shifted, added, etc.

W - Displacement of a ship.

θ - Angle of heel.

δ - Density.

Δ - Displacement at summer draft.

FORMULAE

Angle of Heel, or Loll.

Due to weight out of centre-line—$\cot\theta = \dfrac{W \times GM}{w \times d}$

Due to negative GM—$\tan\theta = \sqrt{\dfrac{2GM}{BM}}$

Areas of Common Figures.

Circle - - -	Area	$= \pi r^2$
Rectangle - -	,,	$= ab$
Square - - -	,,	$= a^2$
Trapezoid - -	,,	$= \dfrac{h}{2}(a + b)$
Triangle - -	,,	$= \dfrac{bh}{2}$
,, - -	,,	$= \dfrac{ab \times \sin\theta}{2}$
,, - -	,,	$= \sqrt{s(s - a)(s - b)(s - c)}$

Areas of Waterplanes, etc.

Five-eight rule -	Area	$= \dfrac{h}{12}(5x + 8y - z)$
Simpson's first rule -	,,	$= \dfrac{h}{3}(t + 4u + 2v + 4w + 2x + 4y + z)$
Simpson's second rule	,,	$= \dfrac{3}{8}h(t + 3u + 3v + 2w + 3x + 3y + z)$

B—Shift of B.

$$BB_1 = \frac{v \times gg_1}{V}$$

Bilging.

Sinkage due to empty compartment $= \dfrac{v}{A - a}$

Sinkage due to compartment with cargo $= \dfrac{vp}{A - ap}$

Sinkage, if W/T flat exists in compartment $= \frac{vp}{A}$

Change of trim $= \frac{W \times BB_1}{\text{M.C.T.1C.}}$

$BB_1 = \frac{v \times d}{V}$

BM.

For all shapes - - - $BM = \frac{I}{V}$

For box shapes only - - $BM = \frac{b^2}{12D}$

Approximate, for ship shapes $BM = \frac{ab^2}{D}$

BM_L

For all shapes - - - $BM_L = \frac{I_L}{V}$

For box shapes - - - $BM_L = \frac{l^2}{12D}$

Circumference of a Circle $= 2\pi r$

Density and Draft.

For box shapes - - $\frac{\text{New draft}}{\text{Old draft}} = \frac{\text{Old density}}{\text{New density}}$

For ship shapes - - Sinkage $= \frac{F\,(1025 - \delta)}{25}$

For ship shapes - - Sinkage $= \frac{W\,(\delta - \delta_1)}{\delta_1 \times \text{T.P.C.}}$

Fresh water allowance - - - $= \frac{\Delta}{4T}$ millimetres

Draft.

Loading to keep constant aft $d = \frac{\text{M.C.T.1C.} \times L}{\text{T.P.C.} \times l}$

Mean draft $\sim$ draft at $F = \frac{d \times \text{amount out of designed trim}}{L}$

Drydocking.

New GM = Old $GM - \frac{P \times KM}{W}$

Where P = Old displacement − new displacement

Or $P = \frac{\text{M.C.T.1C.} \times t}{l}$ (approx.)

Dynamical Stability.

(Moseley's formula)—

Dynamical stability $= W\left\{\frac{v(gh + g_1h_1)}{V} - BG \times \text{Versine } \theta\right\}$

Free Surface of Liquids.

Rise of G, due to free surface of any shape - $GG_1 = \frac{i}{V}$

Rise of G, due to rectangular free surface - $GG_1 = \frac{lb^3}{12V}$

G — Shift of G.

$GG_1 = \frac{w \times d}{W}$

GM.

By Inclining Experiment - - $GM = \frac{w \times d}{W} \times \frac{CF}{FL}$

GZ.

At small angles - - $GZ = GM \times \sin \theta$

By Atwood's Formula - $GZ = \frac{v \times hh_1}{V} - BG \times \sin \theta$

By the Wall Sided Formula $GZ = \sin \theta \, (GM + \frac{1}{2} BM \times \tan^2 \theta)$

Moment to Change Trim by 1 Centimetre.

For all shapes - - - - - - M.C.T.1C. $= \frac{W \times GM_L}{100L}$

Layer Correction.

Displacement of layer - $= \frac{\text{T.P.C.} \times d \times \text{Amount out of trim}}{L}$

Moment of Inertia.

Transversely, for rectangular waterplanes - - $I = \frac{lb^3}{12}$

Longitudinally, for rectangular waterplanes - - $I = \frac{bl^3}{12}$

Permeability.

Permeability $= \frac{\text{Stowage factor} - (1 \div \text{Relative density})}{\text{Stowage factor}}$

Pressure.

At depth D - - - - - - - - $= \delta D$ tonnes-m^2

On area A, at depth D - - - - - - $= \delta AD$ tonnes

Sinkage.

Due to bilging an empty compartment - - $X = \dfrac{v}{A - a}$

Due to bilging a compartment with cargo - $X = \dfrac{vp}{A - ap}$

Due to added weights - - Bodily sinkage $= \dfrac{w}{\text{T.P.C.}}$

Statical Stability.

At small angles - Moment of statical stability $= W \times GM \times \sin\theta$

At any angle - Moment of statical stability $= W \times GZ$

Attwood's Formula - Moment of S.S. =

$$W\left\{\frac{v \times hh_1}{V} - BG \times \sin\theta\right\}$$

Surface Areas

Box shape - - - - - - Area $= 2\,(al + bl + ab)$

Cube - - - - - - - Area $= 6a^2$

Cylinder - - - - - - Area $= 2\pi r\,(r + l)$

Sphere - - - - - - Area $= 4\pi r^2$

Tons per Centimetre Immersion.

In water of density δ - - - T.P.C. $= \dfrac{\delta \times A}{100}$

In salt water - - - - T.P.C. $= \dfrac{1{\cdot}025A}{100}$

Trim.

Change of trim $= \dfrac{\text{Moment changing trim}}{\text{M.C.T.1C.}}$

Loading to produce desired trim - M.C.T. = M.C.T.1C. $(t \sim t_1)$

Volumes.

Box shape - - - - - Volume $= abl$

Cube - - - - - - Volume $= a^3$

Cylinder - - - - - Volume $= \pi r^2 l$

Hollow round section - - - Volume $= \pi l(R + r)(R - r)$

Hollow sphere - - - - Volume $= \dfrac{4\pi(R^3 - r^3)}{3}$

Sphere - - - - - - Volume $= \dfrac{4\pi r^3}{3}$

Wedge, or prism - - - - Volume $= Al$

Wetted Surface.

Area $= L\{1{\cdot}7d + (C \times b)\}$

DEFINITIONS

Angle of Vanishing Stability.—The angle at which a ship's stability becomes 0: numerically the same as the range of stability.

Centre of Buoyancy.—The geometrical centre of the underwater part of the ship.

Centre of Flotation.—The point about which the ship heels and trims. The centre of gravity of the waterplane.

Centre of Gravity.—The centre of all the weight in a body. The point about which the body would balance.

Deadweight.—The weight of all cargo, stores, bunkers, etc., in a ship.

Deadweight Moment.—The total moment about the base line of all the components of deadweight (cargo, stores, fuel, etc.).

Displacement.—The actual weight of the ship and all aboard her at any time.

Draft.—The depth of the bottom of the keel below the surface of the water. Measured forward and aft.

Draft Mean.—The mean of the drafts fore and aft, as it would be measured amidships.

Draft at F.—The ship's draft, measured at the position of the centre of flotation.

Dynamical Stability.—The amount of work done in inclining a ship to any given angle of heel.

Equilibrium.—The state of balance of a body.

Fluid KG or GM.—The ship's KG or GM, after allowance has been made for the effect of free surface of liquids.

Force.—Any push or pull exerted on a body.

Freeboard.—The distance from the deck line to the water.

Fresh Water Allowance.—The amount by which the ship would increase her draft on passing from salt to fresh water.

Height of the Metacentre.—The height of the metacentre above the keel

Inertia.—The resistance of a body to motion or to change of motion.

Initial Stability.—The statical stability of a ship at a small angle of heel. Indicated by GM.

Isochronous Rolling.—The name given to the rolling of a ship when the period of each roll is exactly the same.

KG.—The height of the centre of gravity above the keel.

KN.—The righting lever which the ship would have if G were placed at the keel.

Law of Archimedes.—A body immersed in a liquid appears to suffer a loss in weight equal to the weight of liquid which it displaces. From this law we conclude that a floating body displaces its own weight of water.

Light Displacement.—The displacement of a ship when she is floating at her designed light draft. The weight of the hull, machinery, spare parts and water in the boilers.

Loaded Displacement.—The displacement of a ship when she is floating at her designed summer draft. The light displacement plus the deadweight.

Longitudinal Metacentric Height.—The height of the longitudinal metacentre above the centre of gravity.

Mean Draft.—The mean of the ship's drafts fore and aft.

Metacentre.—The point at which the vertical line through the centre of buoyancy, at a small angle of heel, cuts the ship's centre-line. It is only considered to exist for angles of heel of up to about 15°.

Metacentric Height.—The height of the transverse metacentre above the centre of gravity.

Moment.—The attempt of a force to turn a body. It is usually measured by the product of the force and the length of lever.

Moment to Change Trim by One Centimetre.—The moment to change the ship's trim by one centimetre.

Moment of Statical Stability.—The moment which will try to return a ship to the upright when she is inclined.

Period of a Ship.—The time taken by a ship to roll from one side to the other and back again.

Period of Waves.—The interval between the passages of any two consecutive wave crests.

Prismatic Bodies.—The term "Prismatic" is used in stability to indicate a body which has a constant cross-section throughout its length. For example in the case of a box-shaped vessel which is on an even keel fore and aft, but heeled, the immersed and emerged wedges are prismatic.

The volume of a prismatic body is the area of its cross-section multiplied by its length.

Range of Stability.—The angular range over which a ship will have positive statical stability. The angle to which the ship could heel before she would tend to capsize.

Reserve Buoyancy.—The volume of a ship's hull between the waterplane and the freeboard deck.

Righting Lever.—The perpendicular distance between the centre of gravity and the vertical line through the centre of buoyancy. The lever on the ends of which the weight of the ship acts to return her to the upright when she is heeled.

Solid KG or GM.—The ship's KG or GM, neglecting the effect of free surface of liquids.

Stiff Ship.—A ship which has a large moment of statical stability. One having a large metacentric height, or righting lever.

Synchronism.—Said to occur when the ship's period of roll is the same as that of the waves.

Tender Ship.—A ship which has a small moment of statical stability. One having a small metacentric height, or righting lever.

Tonnages.—Measures of certain spaces in a ship, expressed in terms of 100 cubic feet to the "ton".

Tonnes per Centimetre Immersion.—The weight which must be added to a ship in order to cause her to sink one centimetre bodily.

Trim.—The difference of the drafts forward and aft. The longitudinal equivalent of heel.

PROBLEMS

Increase of Pressure with Depth

1. A flat plate, 5 metres long and 2 metres wide, is placed horizontally at a depth of 12 metres below the sea surface. Find the pressure per square metre and the total pressure on the plate.

2. A sealed box is made of a metal which is capable of withstanding a pressure of 500 grammes per square centimetre. To what depth, in salt water, can the box be sunk before it will collapse?

3. Find the total pressure, in tonnes, on a keel plate which has an area of 82 m^2 and which is 9·0 m below the sea surface.

4. Find the average pressure, in kilogrammes per square centimetre, on the body of a diver who is working in salt water at a depth of 17 metres.

5. A tank is being tested and is pressed-up with salt water until it overflows from the air pipe. If the top of the pipe is 8·6 metres above the tank-top, find the pressure on the tank-top in kg/cm^2.

6. A double bottom tank is 1·6 metres deep and contains oil of relative density 0·895. If a sounding gives a reading of 4·8 metres, what is the pressure on the tank top in kg/cm^2?

Answers—

1. 12·3 tonnes; 123 tonnes.
2. 4·878 m.
3. 756·5 tonnes.
4. 1·742 kg.
5. 0·882 kg/cm^3.
6. 0·286 kg/cm^2.

Floating Bodies and Density

7. A log weighs 4 tonnes and floats in water of density 1·024 $tonnes/m^3$. What volume of water does it displace?

8. A floating body displaces 21·45 cm^3 of water of density 1·010 $tonnes/m^3$. Find its weight in grammes.

9. A ship displaces 2941 m^3 of water of density 1022 kg/m^3. Find the density of the water in which she would displace a volume of 2945 m^3.

10. A log is 10 × 1 × 0·7 metres and weighs 4·08 tonnes. How much of it would be above the surface in water of 1·020 gr/cm^3?

11. A block of hardwood, 4·0 m × 0·8 m × 0·3 m, weighs 985 kilogrammes. Will it float or sink if it is placed in water of relative density 1·026?

Answers—

7. 3·906 m^3.
8. 21·66 gr.
9. 1·0206.
10. 3·0 m^3.
11. Block has same density as the water —in theory it neither floats nor sinks.

Loadlines

12. A ship is 120 metres long. At her summer draft of 7·626 metres, her freeboard is 2186 mm., her displacement is 10720 t and her T.P.C. is 16·58 t. Find the distances between her load lines; also draw the lines.

13. Find the distances between the load lines of a vessel which is 60 metres Long. At her summer draft of 4·053 metres, her displacement is 1861 t and her T.P.C. is 4·67 t.

14. A ship, which is marked with lumber load lines is 135 metres long and her summer draft is 8·042 m. Find the distances between: (*a*) the *LS* and *LT* lines; (*b*) the *LS* and *LW* lines; (*c*) Where would the *LWNA* line be placed?

Answers—

12. *S* to *T* and *S* to *W*: 159 mm.
    S to *F* and *T* to *TF*: 162 mm.
13. *S* to *T* and *S* to *W*: 84 mm.
    S to *F* and *T* to *TF*: 100 mm.
    W to *WNA*: 50 mm.
14. *LS* to *LT*: 168 mm.
    LS to *LW*: 223 mm.
    LWNA level with ordinary *W* line.

Areas

15. Find the area of a square whose sides are 11 cm long.

16. A square has an area of 43 m². Find the length of its sides.

17. What is the area of a rectangle which has sides of 2·6 cm and 11·5 cm?

18. A box-shaped ship is 172 m long and 27 m wide. Find the area of its waterplane when it is upright and on an even keel.

19. What would be the area of the waterplane of the ship in the last question if she were on an even keel but had a list of 20°?

20. A triangle has a vertical height of 6·2 cm and a base of 8·8 cm. Find its area.

21. Find the area of a triangle having two of its sides 3·8 cm and 9·5 cm long, the angle between them being 36°.

22. A triangular plate has sides of 10·7 m, 16·5 m, and 24·0 m. What is its area?

23. Find the area and circumference of a circle of radius 2·62 metres.

24. A circle has a circumference of 32·8 cm. Find its area.

25. A circle has an area of 57·6 cm². Find its circumference.

Answers—

15. 121 cm².
16. 6·36 m.
17. 29·9 cm².
18. 4644 m².
19. 4942·1 m².
20. 27·28 cm².
21. 10·613 cm².
22. 74·5 cm².
23. 21·57 m²; 16·47 m.
24. 85·58 cm².
25. 26·91 cm.

Surface Areas and Volumes

26. Find the surface area and volume of a cube which has edges of 3 cm length.

27. A cube is made of a metal which has a density of 7·290 gr/cm³. Its edges are 2·4 cm long. Find its weight.

28. Find the underwater volume and displacement, in sea water, of a box-shaped vessel of 120 m long, 17 m beam and 3·00 m draft.

29. A box-shaped lighter is 25·00 m long, 6·00 m wide and has a light draft of 0·60 m. How many tonnes of cargo must she load, when floating in salt water, in order to increase her draft to 2·00 m?

30. Find the weight of a log of wood, 10 metres long, and of cross-section 90 centimetres by 40 centimetres, which has a relative density of 0·750.

31. A box-shaped lighter is 35 m long, 8 m beam and weighs 420 tonnes. Find her draft in salt water.

32. A box-shaped lighter is 25·0 m long, 6·5 m wide and floats at a draft of 80 cm in salt water. What is her weight?

33. Find the surface area of a rectangular tank of 9·0 × 3·0 × 1·5 metres.

34. Find the volume of a prismatic-shaped lighter, if the area of each end is 35 m^2 and its length is 50 m.

35. Find the volume and surface area of a sphere of radius 2·1 cm.

36. A sphere of radius 21 centimetres weighs 200 kilogrammes in air. What will be its apparent weight when immersed in sea water?

37. A hollow sphere has an internal radius of 3 cm and an external radius of 6 cm. Find the volume of material in it.

38. Find the weight of a hollow sphere of internal radius 6 cm, external radius 10 cm and relative density 2·700.

39. Find the volume and surface area of a cylinder, 80 cm long and having a radius of 14 cm.

40. A cylindrical boiler has a diameter of 3·0 m, a length of 4·5 m and weighs 80 tonnes. Would it be possible, by sealing the openings in it, to float it ashore in an out-of-the-way port?

41. Find the weight, in tonnes, of a hollow mast which is 10 m long and 50 cm outside diameter, made of steel 20 mm thick. Assume that the steel weighs 8·00 tonnes/m^3.

Answers—

26. 54 cm^2; 27 cm^3.
27. 100·8 gr.
28. 6120 m^3; 6273 tonnes.
29. 215·25 tonnes.
30. 2700 kg.
31. 146·3 cm.
32. 133·25 tonnes.
33. 90 m^2.
34. 1750 m^3.
35. 38·81 cm^3; 55·44 cm^2.
36. 160·2 kg.
37. 792 cm^3.
38. 8·869 kg.
39. 49267 cm^3; 8271 cm^2.
40. Yes. Weight of boiler is 30 tonnes. Weight of equal volume of salt water is 32·6 tonnes.
41. 2·413 tonnes.

Simpson's Rules, Etc.

42. Use Simpson's First and Second Rules to find the area of part of a waterplane which has the following ordinates, spaced 6·0 metres apart:—

2·4, 3·7, 5·3, 6·6, 7·6, 8·1, 8·2 metres.

43. Use Simpson's First Rule to find the area of part of a waterplane, which has a common interval of 3·0 metres and ordinates of:—

1·3, 3·5, 4·8, 5·4, 5·6, 5·0, 3·9 metres.

44. Find the area of a waterplane which has the following half-ordinates and a common interval of 8 metres:—

0, 3·0, 4·7, 6·2, 7·3, 7·8, 7·3, 6·9, 5·3, 3·1, 0 metres.

45. Find the area of a waterplane which has the following ordinates and a common interval of 6 metres:—

0, 5·4, 8·2, 9·0, 8·6, 7·6, 5·4, 2·1, 0 metres.

46. Part of a waterplane has the following half-ordinates and a common interval of 4·0 metres:—

1·1, 2·2, 3·7, 5·3, 6·9, 8·2, 8·7, 9·0, 8·9, 8·8 metres.

Find its area.

47. A prismatic-shaped tank is 15 metres long. Its end is divided into ordinates of 5·0, 4·5, 3·8 and 2·0 metres, spaced 2·0 metres apart. Find the area of the end and the volume of the tank.

48. Find the area of a waterplane which has the following ordinates and a common interval of 5 metres:—

0·2, 3·6, 6·7, 8·9, 10·7, 12·0, 12·7, 12·0, 11·4, 8·8, 5·0, 0·3 metres.

49. Find the area of a bulkhead which has the following ordinates, spaced 2·0 metres apart:—

3·0, 4·7, 7·4, 9·5, 11·1, 12·5, 13·6, 14·4 metres.

50. Three ordinates are spaced 12 metres apart and have lengths of 7·4, 11·7 and 17·5 metres. Find the areas between the first and second; also between the second and third ordinates respectively. How does the total area so found compare with that found by Simpson's First Rule?

51. Find the areas between two ordinates, 10 metres apart, which have lengths of 10·3 and 15·0 metres, if the next ordinate has a length of 19·6 metres.

52. Find the coefficient of fineness of a waterplane which has the following half-ordinates, spaced 10 metres apart:—

0·5, 4·0, 7·2, 9·1, 10·0, 9·8, 8·1, 5·1, 0 metres.

The greatest breadth of the waterplane is 20 metres.

53. The midship section of a boat is 3·0 metres deep and is divided horizontally into the following equally spaced ordinates:—

7·5, 7·2, 6·8, 6·0, 5·3, 3·3, 0·5 metres.

Find the coefficient of fineness of the midship section.

54. A ship is divided up into a number of waterplanes, spaced 0·6 metres apart and having the following areas:—

Keel - - -	120	square metres
Waterplane *A* -	1092	,, ,,
,, *B* -	1242	,, ,,
,, *C* -	1355	,, ,,
,, *D* -	1439	,, ,,
,, *E* -	1499	,, ,,
,, *F* -	1548	,, ,,
,, *G* -	1593	,, ,,

D is the light waterplane and *G* the load waterplane. Find:—

(*a*) the light displacement;
(*b*) the loaded displacement;
(*c*) the block coefficient of fineness, assuming the length and breadth of the ship on waterplane *G* to be 130 metres and 17 metres, respectively.

55. The midship section of a ship has the following ordinates, spaced 3 metres apart, below her waterplane:—

16·4, 16·0, 15·5, 15·0, 14·8, 14·0, 12·8, 10·9, 7·9, 1·0 metres.

Find the area of the midship section. Also, if the length of the ship is 130 metres and her displacement is 12,095 tonnes, find her prismatic coefficient of fineness.

56. A waterplane has ordinates, spaced 9 metres apart, of 0·6, 8·4, 14·5, 17·2, 17·5, 17·0, 15·9 and 12·5 metres. There is an intermediate ordinate, midway between the first two, of 5·1 metres. Find the area of the waterplane.

57. Find the area of a waterplane which has ordinates, 7·0 metres apart of:—
12·0, 11·7, 10·9, 9·5, 6·8, 1·8 metres.
There is an intermediate ordinate of 5·4 metres between the last two.

58. A waterplane has ordinates, from forward to aft, of 0·2, 8·0, 12·3, 12·8, 13·1, 10·2, 7·5, 4·0, 0·6 metres, spaced 10 metres apart. There are two intermediate ordinates: one, midway between the two forward ones, is 4·2 metres long; the other, midway between the two after ones, is 2·0 metres long. Find the area of the waterplane.

59. A vessel has waterplanes, 1 metre apart, from the keel upwards, of areas 152, 402, 625, 807, 950 and 1032 square metres. There is also an intermediate ordinate, 50 centimetres above the keel, of 261 square metres. Find the vessel's displacement.

Answers—

42. First rule, 220·0 m².
Second rule, 220·3 m².
43. 81·6 m².
44. 838·4 m².
45. 281·6 m².
46. 463·8 m².
47. 23·9 m²; 358·9 m³.
48. 464·6 m².
49. 135·1 m².
50. 113·1 m²; 173·7 m²; the same;
286·8 m².
51. 126·6 m².
52. 0·68.
53. 0·73.
54. Light displacement, 2,835·4 tonnes.
Load displacement, 5,642·8 tonnes.
Block coefficient, 0·5930.
55. 116·3 m³; 0·780.
56. 886·8 m².
57. 327·8 m².
58. 694·7 m².
59. 3467 tonnes.

Forces, Moments, Etc.

60. A weight of 3 tonnes is placed on a beam, so that its centre of gravity is 25 metres from the end. What is the moment about this end?

61. A bar is pivoted in the middle and a man pushes on it, in a clockwise direction, with a force of 25 kilogrammes at a distance of 300 centimetres from the middle. Another man, on the other side, also pushes in a clockwise direction with a force of 20 kilogrammes. If the second man is 250 centimetres from the middle of the bar, find the moment about its centre.

62. If, in the above question, the second man turned and pushed with the same force in an anticlockwise direction, what would be the moment?

63. A man presses down on the longer end of a bar with a force of 50 kilogrammes. The bar is 7 metres long and is supported at a distance of 2 metres from its end. How much weight can the man lift on the shorter end?

64. Four men, working at a capstan, each push on the bars at a distance of 3·0 metres from the centre. If they push with forces of 20, 25, 30 and 35 kilogrammes, respectively, find the moment to turn the capstan.

65. If a rope is wound around the capstan in the last question, and the radius from the centre of the capstan to the centre of the rope is 40 centimetres, find the pull on the rope.

66. A wire, wound around a capstan, supports a weight of 0·50 tonnes. The diameter of the capstan barrel is 60 cm. How much force must be exerted by each of four men, each at a distance of 220 cm from the centre, in order to lift the weight?

67. Two weights are placed on a beam: one of 2 tonnes at a distance of 10 metres from one end; the other of 3 tonnes at a distance of 7 metres from the same end. What is the moment about the end of the beam?

68. A weight of 150 kilogrammes is placed on a see-saw at a distance of 400 cm from the middle. What weight must be placed on the other side, at a distance of 300 cm from the middle, in order to balance the seesaw?

Answers—

60. 75 tonne/metres.
61. 12500 kg/cm.
62. 2500 kg/cm.
63. 125 kg.
64. 330 kg/metres.
65. 825 kg.
66. 34·1 kg.
67. 41 tonne/metres.
68. 200 kg.

Centre of Gravity (General)

69. A plank weighs 62 kilogrammes. What will be the shift of its centre of gravity if a weight of 38 kilogrammes is placed on it at a distance of 400 cm from its original centre of gravity?

70. A weight of 0·400 tonnes is added to a body which weighs 2·00 tonnes, at a distance of 120 cm from its centre of gravity. Find the shift of *G*.

71. A body, weighing 32 tonnes, has a weight of 8 tonnes removed from it. What is the shift of *G*, if the centre of gravity of the weight removed was 5·00 metres from that of the body?

72. A beam carries a weight of 4·8 tonnes at a distance of 6·0 metres from one end and the centre of gravity of the whole mass is 20 metres from that end. If the weight of the beam alone is 8·4 tonnes, find the distance of the centre of gravity of the beam from the end mentioned.

73. A loaded boat weighs 6 tonnes. If a weight of 300 kilogrammes is shifted from one end to the other, through a distance of 20 metres, find the shift of *G*.

74. A boat weighs 1·512 tonnes and when a man, weighing 70 kilogrammes, sits in the bottom, the centre of gravity is 80 centimetres above the keel. What will be the new height of the centre of gravity if the man stands up so as to raise his weight by 160 centimetres?

75. A see-saw has a number of weights placed on one end at a distance of 150 cm from the fulcrum. The centre of gravity of the whole mass is 30 cm from the fulcrum and the total weight is 270 kilogrammes. A number of the weights are then moved for a distance of 300 cm, along the see-saw to the other end, so that it balances. Find the amount of weight shifted.

76. A table-top has a number of weights on it and the whole mass weighs 275 kilogrammes. How far would it be necessary to shift a weight of 25 kilogrammes in order to cause the centre of gravity of the mass to shift for one centimetre?

77. Find the position of the centre of gravity of the waterplane in Question 44.

78. Find the position of the centre of gravity of the waterplane in Question 45.

79. A waterplane has the following ordinates spaced 10 metres apart, from, forward to aft:—

0·6, 8·4, 14·9, 18·6, 20·0, 19·3, 16·1, 10·7, 2·5 metres.

Find the position of its centre of gravity relative to the mid-ordinate.

80. A waterplane has ordinates, spaced 20 metres apart, from forward to aft, of:—

0·2, 2·8, 6·9, 9·6, 10·3, 10·7, 9·8, 7·7, 3·2 metres.

There is also an appendage of 20 square metres, with its centre of gravity 2·0 metres abaft the aftermost ordinate. Find the position of its centre of gravity.

81. A waterplane is 60 metres long and has ordinates of 0·4, 6·6, 10·1, 14·3, 11·8, 8·2 and 3·5 metres, from forward to aft. There is also an appendage of 12·4 square metres, with its centre of gravity 32·2 metres abaft the mid-ordinate. Find the position of the centre of gravity.

82. Find the height of B in a ship which has waterplanes 1·0 metre apart, from the keel upwards, of 110, 1071, 1203, 1331, 1412, 1486, 1525 square metres.

83. A small vessel, 48 metres long, has the following underwater cross-sections, from forward to aft:—

0·9, 11·7, 17·3, 24·2, 29·5, 26·0, 20·4, 12·8, 3·4 square metres.

Find the longitudinal position of B.

84. A ship has the following underwater cross-sections, 12 metres apart, from forward to aft:—

4, 31, 59, 78, 80, 82, 81, 79, 63, 45, 20 square metres.

There is also an appendage of 384 cubic metres, with its centre of gravity 70 metres abaft the mid-ordinate. Find the position of B.

85. A yacht has waterplanes, 20 cm apart, from the base line (a horizontal line through the top of the keel amidships) to the waterplane as follows:—

6·3, 9·7, 11·9, 13·6, 14·5 square metres.

In addition, the ballast keel forms an appendage of 0·75 cubic metres, with its centre of gravity 80 centimetres below the base line. Find the position of B relative to the base line.

86. A waterplane has ordinates, 8 metres apart, from forward to aft, of 0·2, 4·8, 8·6, 12·6, 13·4, 12·2, 9·7, 4·0 metres. There is also an intermediate ordinate of 2·1 metres, midway between the two foremost ordinates. Find the area of the waterplane and the position of its centre of gravity.

87. A waterplane, 120 metres long, is divided into eight equal parts, giving half-ordinates, from forward to aft of 0·0, 5·4, 9·0, 10·7, 11·0, 10·8, 9·5, 6·8, 1·2 metres. There are also two intermediate half-ordinates: one of 3·8 metres, between the foremost two; one of 5·0 metres between the aftermost two. Find the area of the waterplane and the position of its centre of gravity.

88. A ship, floating at her light draft, has waterplanes 1 metre apart, from the keel upwards, of 270, 2748, 3020, and 3231 square metres. There is also an intermediate waterplane, 0·5 metre above the keel, of 2228 square metres. Find the ship's displacement and KB at her light draft.

Answers—

69. 152 cm.
70. 20 cm.
71. 166·7 cm.
72. 28 m.
73. 1 m.
74. 83·7 cm.
75. 27 kg.
76. 11 cm.
77. 0·55 m from mid-ordinate.
78. 2·46 m from mid-ordinate.
79. 1·28 m abaft mid-ordinate.
80. 16·2 m abaft mid-ordinate.
81. 2·21 m abaft mid-ordinate.
82. 3·36 m above keel.
83. 0·59 m abaft mid-ordinate.
84. 5·54 m abaft mid-ordinate.
85. 35 cm above base line.
86. 511·7 m^2; 31·3 m from forward.
87. 987·5 m^2; 1·62 m abaft mid-ordinate
88. 8208 tonnes: 1·67 m above keel.

Moment of Inertia

89. A rectangular surface is 12 metres long and 5 metres wide. Find its moment of inertia about (*a*) the transverse centre-line; (*b*) the longitudinal centre-line.

90. A box-shaped lighter is 120 metes long and 20 metres beam. Find the moments of inertia of its waterplane about its longitudinal and transverse centre-lines.

91. Find the moments of inertia of a rectangle, 6 metres long and 2 metres wide. about its centre-lines.

Answers—

89. (*a*) 125; (*b*) 720.
90. 80,000; 2,880,000.
91. 4; 36.

The Effect of Density on Draft

92. A box-shaped barge floats at a draft of 2·55 m in water of density 1·004 tonnes/m^3. What would be her draft in water of density 1·020?

93. A ship has a fresh water allowance of 175 mm. To what depth could she submerge her loadline when loading in dock water of density 1011 kg/m^3?

94. A ship has a fresh water allowance of 185 mm and a summer draft of 8·52 m. Find the draft to which she may load in river water of density 1·007 tonnes/m^3.

95. A box-shaped vessel has a draft of 4·90 m in salt water. On entering a dock, her draft becomes 5·00 m. Find the density of the dock water.

96. A ship has a fresh water allowance of 175 mm. By how much will she change her draft if she passes from water of density 1004 to water of density 1021 kg/m^3?

97. A box-shaped barge has a draft of 1·12 m in salt water, when she is fully loaded. To what draft could she load in water of 1·015 tonnes/m^3?

98. A box-shaped lighter draws 1·95 m in water of density 1004 kg/m^3. Find her draft in water of density 1020 kg/m^3.

99. To what depth can a ship submerge her load-line in dock water of density 1012 kg/m^3, if her fresh water allowance is 150 mm?

100. A ship enters port with a salt water draft of 7·18 m. If her fresh water allowance is 122 mm, what will be her draft in a dock where the relative density of the water is 1·009?

101. A box-shaped vessel has a draft of 3·31 m in salt water. Find her draft in water of density 1·000 tonnes/m^3.

102. A box-shaped lighter draws 4·10 m in water of relative density 1·010. What will be her draft in sea water?

103. By how much can a ship submerge her load-line in water of density 1011 kg/m^3, if her fresh water allowance is 202 mm?

Answers—

92. 2·51 m.
93. 98 mm.
94. 8·65 m.
95. 1·005 tonnes/m^3.
96. 119 mm.
97. 1·13 m.
98. 1·92 m.
99. 78 mm.
100. 7·26 m.
101. 3·39 m.
102. 4·04 m.
103. 113 mm.

T.P.C.

104. Find the tonnes per centimetre immersion of a box-shaped vessel, 70 metres long and 10 metres beam.

105. A ship is 150 metres long, 16 metres beam and floats at a draft of 5·00 metres. If the coefficient of fineness of the waterplane at that draft is 0·763, find her T.P.C.

106. A box-shaped lighter is 25 metres long, 6 metres wide and floats at a draft of 1·10 metres fore and aft. What is her T.P.C. and what will be her new draft after 30 tonnes of pig-iron have been spread evenly over the bottom?

107. At a given draft, a ship of 120 metres length and 15 metres beam has a coefficient of fineness of the waterplane of 0·770. Find her T.P.C. at this draft.

108. A vessel's waterplane has a common interval of 6·0 metres and ordinates of 0, 4·4, 7·4, 9·7, 10·3, 10·0, 9·8, 5·5, 0·6 metres. Find her T.P.C.

109. A ship's waterplane has ordinates of 0·5, 3·8, 5·9, 6·5, 6·8, 6·6, 5·7, 4·4, and 2·5 metres, spaced 12 metres apart. Find her T.P.C. in fresh water.

110. A waterplane has an area of 1960 square metres. Find the T.P.C. in water of density 1·012 t/m^3.

111. A vessel has a T.P.C. of 18·50 t in salt water. What would the T.P.C. be in water of density 1·010 t/m^3?

112. What would be the effect of loading a weight of 140 tonnes, in a ship which has a T.P.C. of 10 t?

113. A ship floats at a mean draft of 6·25 metres. If her T.P.C. is 12·0 t, what would be her new mean draft after 300 tonnes of cargo have been loaded?

114. Find the bodily rise of a ship which has a T.P.C. of 7·0, if a weight of 120 tonnes is discharged.

115. A ship is loading in dock water of density 1·012 t/m^3 and has a mean draft of 7·16 metres. Her T.P.C. in salt water is 17·82 t. Find the new mean draft after she has loaded a further 264 tonnes.

116. A ship has a mean draft of 4·38 metres in water of relative density 1·015. Her T.P.C. in salt water is 15·20 t. What will be her new draft after she has discharged 360 tonnes of water ballast?

Answers—

104. 7·2 t.
105. 18·8 t.
106. 1·5 t; 1·3 m.
107. 14·2 t.
108. 3·57 t.
109. 10·08 t.
110. 19·84 t.
111. 18·23 t.
112. Ship sinks 14 cm bodily.
113. 6·50 m.
114. 17 cm.
115. 7·31 m.
116. 4·14 m.

Loading to a Given Loadline

117. A vessel has a summer draft of 6·730 metres, a F.W.A. of 152 mm. and T.P.C. of 21·86 t. She is loading in a dock, where the density of the water is 1·013 t/m^3 and her present mean draft is 6·64 metres. How much more cargo can she load in order to be at her summer load line on reaching salt water?

118. A ship is loading in an upriver port, where the relative density of the water is 1·008 and her present mean draft is 6·98 metres. Her summer draft is 7·165 metres, F.W.A. is 116 mm., and her T.P.C. is 17·91 t. Calculate how much more cargo she can load in order to float at her summer load line on reaching salt water, if she expects to use 32 tonnes of fuel and stores on the way downriver.

119. A ship has a summer freeboard of 1764 mm., T.P.C. of 19·54, and F.W.A. of 135 mm. She is loading in dock water of density 1·015 t/m³ and her present freeboards are 1792 mm. on the port side and 1816 mm. on the starboard side. How much more cargo can she load in order to float at her summer load line on entering salt water?

120. A vessel arrives off a port, in salt water, with an even-keel draft of 7·91 metres. Her F.W.A. is 162 mm. and her T.P.C. is 23·04 t. She is to discharge cargo into lighters in order to enter a dock, where the density of the water is 1·016 t/m³. The depth on the dock sill is 8·10 metres, and she is to cross it with a clearance of 30 cm. Assuming that the ship remains on an even keel, find the least amount of cargo to discharge into the lighters.

121. A ship is loading in dock water of relative density 1·020. The upper edge of her summer load line is level with the water on the port side: whilst the lower edge of her summer load line is 6·5 cm. above water on the starboard side. Her F.W.A. is 154 mm. and her T.P.C. is 22·98 t. How much more can she load in order to float at her summer load line in salt water?

122. A ship has a summer draft of 8·094 m., her T.P.C. is 23·25 t, and her displacement at summer draft is 15250 t. She is loading in dock water of relative density 1·010 and her present mean draft is 8·22 metres. How much more can she load in order to float at her tropical load line on reaching salt water?

123. A vessel is loading at a berth in river water of density 1·009 t/m³. Her summer freeboard is 1948 mm., T.P.C. is 23·08 t, and F.W.A. is 164 mm. Her present freeboards are 2077 mm. on the port side and 2053 mm on the starboard side. On completion of loading, she is to proceed to the river mouth, using 12 tonnes of fuel and stores on the way, and is then to load 240 tonnes of cargo from lighters before proceeding to sea. How much more cargo can she load in the upriver berth, in order to be at her summer load line in salt water on leaving the river mouth?

124. A wall-sided ship has a waterplane area of 2175 square metres and displaces 12800 tonnes when floating at her summer draft of 7·200 metres. She is floating in water of relative density 1·015 and the lower edge of her T.F. load line is 3 cm. above water on the port side, whilst the lower edge of the summer load line is level with the water on the starboard side. How much cargo can she load in order to float at her Tropical load line in salt water?

125. A vessel arrives at a port at a river mouth in water of density 1·022 t/m³, floating at an even-keel draft of 8·06 metres (Freeboard, 2·47 metres). She is to discharge as much cargo as possible and then to proceed to an upriver port, where the relative density of the water is 1·012 and where there is a bridge under which she must pass. The bridge is 23·00 metres above water level; the truck of the ship's mast is 19·40 metres above the freeboard deck, and must have a clearance of 1·0 metres for passing under the bridge. F.W.A. is 125 mm. and the T.P.C. is 21·04 t in salt water. If the vessel will burn 9½ tonnes of fuel on the way upriver, and assuming the ship to remain on an even keel throughout, find the maximum amount of cargo that she can discharge at the river mouth.

Answers—

117. 352 t.
118. 497 t.
119. 182 t.
120. 387 t.
121. 174 t.
122. 323 t.
123. 276 t.
124. 223 t.
125. 368 t.

Shift of G in Ships

126. A ship has a displacement of 2000 tonnes. Find the shift of her centre of gravity if a weight of 100 tonnes is shifted 12 metres across a hold.

127. A weight of 500 tonnes is loaded into a ship so that its centre of gravity is 10 metres from that of the ship. Find the shift of *G* if the ship's original displacement was 3000 tonnes.

128. A ship and her cargo displace 7200 tonnes. What will be the shift of the centre of gravity if a weight of 80 tonnes is removed from a point 100 metres from the original centre of gravity of the ship?

129. Find the effect of adding a weight of 80 tonnes at a distance of 120 metres from *G* in a ship, the original displacement of which was 7600 tonnes.

130. What will be the shift of *G* in a ship of 3600 tonnes displacement, if a weight of 40 tonnes is moved 16 metres across the deck?

131. A ship has a displacement of 11,000 tonnes. Calculate the shift of *G* if a weight of 1000 tonnes is removed from a point 60 metres from the original centre of gravity.

132. A ship has a displacement of 3600 tonnes and a *KG* of 3·12 metres. A weight of 60 tonnes is raised from a hold into a 'tween deck, through a vertical distance of 7·2 metres. Find the new *KG*.

133. A weight of 250 tonnes is loaded into a ship at a height of 6·6 metres above her centre of gravity. If her original *KG* was 4·75 metres and her new displacement, after the weight has been loaded, is 2200 tonnes, find the new *KG*.

What would the *KG* have been if the weight had been loaded at a distance of 6·6 metres below the ship's centre of gravity, instead of above it?

134. A ship has a displacement of 2550 tonnes and a *KG* of 7·40 metres. What would be the *KG* after 950 tonnes of cargo had been loaded, with its centre of gravity 2·50 metres above the keel?

135. Find the shift of *G* and the new *KG* if a weight of 40 tonnes is discharged from a point 1·20 metres below the centre of gravity of a lighter? The lighter's original displacement and *KG* were 680 tonnes and 3·00 metres respectively.

136. A ship's displacement is 2800 tonnes and her *KG* is 4·15 metres. What will be the new *KG* if a weight of 35 tonnes is lowered vertically downwards into the hold for a distance of 12·0 metres?

137. A weight of 9 tonnes is lifted from a hold by means of a derrick, the head of which is 20 metres above the original position of the centre of gravity of the weight. If the ship's displacement is 2700 tonnes, what will be the shift of *G*?

138. A ship has a *KG* of 5·00 metres and a displacement of 3000 tonnes. A weight of 20 tonnes is lifted from the lower hold and placed on deck by means of a derrick, the head of which is 25 metres above the keel. The centre of gravity of the weight was 1·0 metres above the keel when in the hold and 11·5 metres above the keel when on deck. Find:—

(*a*) the *KG* when the weight is hanging on the derrick;

(*b*) the *KG* when the weight has been landed on deck.

139. Fifteen tonnes is lifted by a derrick, the head of which is 15 metres above the original position of the weight, in a ship of 1250 tonnes displacement and *KG* 3·10 metres. What will be the new *KG* when the weight has been lifted through (*a*) 2 metres; (*b*) 12 metres?

140. A ship displaces 2415 tonnes and has a *KG* of 4·50 metres. A weight of 35 tonnes is lifted from the shore by a derrick, the head of which is 20 metres above the keel and is placed in the hold, the final position of its centre of gravity being 3 metres above the keel. Find the *KG* (*a*) when the weight is hanging on the derrick, and (*b*) when the weight has been landed in the hold.

141. A double-bottom tank, when full, has its centre of gravity at a height of 60 cm above the keel and can hold 380 tonnes of water. The *KG* of the ship is 9·40 metres and her displacement is 3700 tonnes when the tank is empty. What will be her *KG* when the tank is filled?

142. A tank holds 252 tonnes of water and its centre of gravity is 44 metres from that of the ship. If the ship's displacement is 3024 tonnes when the tank is full, what will be the shift of *G* caused by pumping it out?

143. 150 tonnes of oil are transferred from a fore peak tank to an after peak tank, the distance between their centres of gravity being 130 metres. Find the shift of *G* due to this, if the ship's displacement is 7500 tonnes.

Answers—

126. 0·60 m.
127. 1·4 m.
128. 1·12 m.
129. 1·2 m towards the c.g. of weight.
130. 0·18 m.
131. 6·0 m.
132. 3·24 m.
133. 5·50 m; 4·00 m.
134. 6·07 m.
135. 7·5 cm; 292·5 cm.
136. 4·00 m.
137. 6·6 cm upwards.
138. (*a*) 5·16 m.
(*b*) 5·07 m.
139. (*a*) 3·28 m.
(*b*) The same.
140. (*a*) 4·72 m.
(*b*) 4·48 m.
141. 8·58 m.
142. 4·0 m.
143. 2·60 m.

KG

144. A ship displaces 2,730 tonnes and has a *KG* of 6·00 metres. She then loads the following weights:—

540 tonnes at 5·0 metres above the keel.
370 tonnes at 8·5 metres above the keel.
110 tonnes at 10·4 metres above the keel.
850 tonnes at 4·6 metres above the keel.

Find her new *KG*.

145. A loaded lighter displaces 856 tonnes and has a *KG* of 1·50 metres. Find the new *KG* after the following weights have been discharged:—

160 tonnes from 2·5 metres above the keel.
40 tonnes from 3·7 metres above the keel.
395 tonnes from 1·2 metres above the keel.

146. A ship leaves port with a displacement of 9060 tonnes and a *KG* of 5·20 metres. During the voyage she consumes the following:—

Oil fuel: 260 tonnes from 0·8 m above the keel.
320 tonnes from 0·7 m above the keel.
Stores: 98 tonnes from 9·5 m above the keel.
Fresh water: 87 tonnes from 10·0 m above the keel.

What will be her *KG* on arrival at her port of destination?

147. The original displacement of a ship was 4285 tonnes and her *KG* was 6·00 metres. Find her new *KG* after she has loaded the following weights:—

800 tonnes at 3.6 metres above the keel.
440 tonnes at 7·0 metres above the keel.
110 tonnes at 5·8 metres above the keel.
630 tonnes at 3·0 metres above the keel.

148. A ship has a *KG* of 6·5 metres and a displacement of 6020 tonnes. Find her *KG* after she has loaded and discharged the following weights:—

Loaded: 500 tonnes at 2·5 metres above the keel.
850 tonnes at 5·0 metres above the keel.
220 tonnes at 8·4 metres above the keel.
Discharged: 300 tonnes from 5·5 metres above the keel.
700 tonnes from 2·6 metres above the keel.

149. Find the new *KG* of a lighter which has loaded and discharged the following weights:—

Discharged: 140 tonnes from 2·5 metres above the keel.
270 tonnes from 1·4 metres above the keel.
Loaded: 215 tonnes at 1·0 metres above the keel.

The original displacement and *KG* were 646 tonnes and 2·00 metres.

150. A ship arrives in port with a *KG* of 6·80 metres and a displacement of 6080 tonnes. Whilst in port, she discharges and loads the following cargo:—

Discharged: 1250 tonnes from 5·0 metres above the keel.
675 tonnes from 3·5 metres above the keel.
420 tonnes from 7·2 metres above the keel.
30 tonnes from 0·7 metres above the keel.
Loaded: 980 tonnes at 3·2 metres above the keel.
550 tonnes at 6·5 metres above the keel.
700 tonnes at 0·6 metres above the keel
70 tonnes at 11·0 metres above the keel.

She then sails on a voyage during which she burns 840 tonnes of oil from 0·6 metres above the keel and uses 60 tonnes of water from 11·0 metres above the keel. Find the *KG*s at the beginning and end of the voyage.

151. The light displacement of a ship is 2875 tonnes. She loads 390 tonnes at 7·0 metres above the keel and 710 tonnes at 2·5 metres above the keel. If her *KG* was then 5·20 metres, what was the light *KG*?

152. A ship displaces 7425 tonnes and has a *KG* of 6·30 metres. She then loads 670 t of cargo at 6·0 metres above the keel and 840 t at 3·0 metres. How much more can she load at 7·0 metres above the keel in order to finish with a *KG* of 6·00 metres?

153. A ship displaces 9500 tonnes and has a *KG* of 5·84 metres. She then loads 550 t of cargo at 4·2 metres above the keel and 720 t at 6·1 metres. She has a further 500 t to load. At what height must this be loaded if the ship is to sail with a *KG* of 5·70 metres?

154. A vessel displaces 4750 t and has a *KG* of 7·05 metres. She then loads 920 t of cargo at 4·5 metres above the keel and 630 t at 7·0 metres: she also discharges 350 t from 8·5 metres above the keel. How much fuel oil can she load into a double bottom tank, at an estimated height of 0·50 metres above the keel, to finish with a *KG* of 6·30 metres: allowing for an estimated rise of *G* of 0·02 metres caused by free surface of the oil?

155. A ship which is completing loading has a *KG* of 7·23 metres and displaces 14600 tonnes. On passage to her next port, she is expected to use 360 t of fuel oil from 0·70 metres above the keel; 40 t of fresh water from 9·40 metres above the keel; and 20 t of stores from 8·8 metres above the keel. Before sailing, she is to load cargo into a tween deck at a height of 9·50 metres above the keel. How much can she load in the tween deck in order to arrive at her next port with a *KG* of 7·45 metres?

156. A ship which is completing loading has a *KG* of 6·82 metres and displaces 11250 t. Before sailing, she has to load deck cargo at an estimated height of 10·5 metres above the keel. On the voyage she is expected to use 180 t of fuel from 5·6 metres above the keel; 30 t of fresh water from 8·8 metres above the keel; and 10 t of stores from 8·2 metres above the keel. How much deck cargo can she load in order to arrive at her next port with a *KG* of 6·95 metres, allowing for a rise of *G* of 0·05 metres due to free surface appearing on voyage?

157. A vessel displaces 9740 t and has a *KG* of 6·06 metres. A 50-tonne lift is to be taken on board by means of a derrick, the head of which will be 25 metres above the keel when lifting. To prevent excessive heel when lifting, the *KG* of the ship must not exceed 6·00 metres when the lift is hanging from the derrick. Find the least amount of cargo to be loaded into a lower hold, at a height of 3·5 metres above the keel, to satisfy this condition.

158. A ship which displaces 7925 t and has a *KG* of 5·42 metres, has to load a deck cargo of timber at a height of 12·0 metres above the keel. On voyage to the next port she expects to use 135 t of fuel and water from the double bottom, from a height of 0·6 metres above the keel; and causing a rise of *G* of 0·04 m, due to free surface effect. Calculate how much timber she can load, allowing 15% of the weight of wood for absorption of water on voyage, to arrive at her destination with a *KG* of 6·10 metres.

Answers—

144. 5·93 m.
145. 1·00 m.
146. 5·41 m.
147. 5·46 m.
148. 6·53 m.
149. 1·73 m.
150. 6·26 m; 7·13 m.
151. 5·62 m.
152. 292 t.
153. 4·11 m.
154. 294 t.
155. 433 t.
156. 204 t.
157. 613 t.
158. 635 t.

BM

159. A box-shaped ship is 120 metres long, 18 metres beam, and floats at a draft of 5·00 metres. What is her *BM*?

160. Find the *BM* of a box-shaped lighter which has a beam of 6·0 metres and floats at a draft of 2·00 metres.

161. What is the height of the metacentre (*KM*) in a box-shaped vessel, of 10·0 metres beam, when floating at a draft of 5·00 metres?

162. Find the *KM* of a box-shaped lighter which has a beam of 7·0 metres and floats at a draft of 1·40 metres.

163. A ship displaces 3860 tonnes and the moment of inertia of her waterplane is 23,350. Find her *KM*, if the *KB* is 3·50 metres.

164. What is the *BM* of a ship of 4160 tonnes, if the moment of inertia of her waterplane is 32,470?

165. Find the GM of a box-shaped lighter, 20 metres long and 6 metres wide, which has a draft of 2·40 metres and a KG of 1·70 metres.

Answers—

159. 3·75 m.
160. 1·50 m.
161. 4·17 m.
162. 3·62 m.
163. 9·70 m.
164. 8·00 m.
165. 0·75 m.

The Inclining Experiment

166. A weight of 25 tonnes is shifted transversely for a distance of 10 metres across the deck of a ship. A plumb-line, which is suspended 4·0 metres above a horizontal batten moves out for a distance of 0·300 metres along the batten. If the ship's displacement is 4950 tonnes, what is her GM, assuming that she was upright at the beginning?

167. When the inclining experiment is performed on a ship of 2304 tonnes displacement, a weight of 15 tonnes is moved for 12 metres across the deck. The plumb-line is 8·0 metres long and moves out 0·43 metres when the ship heels. Find the GM.

168. A ship which has just been completed has a light KM of 10·30 metres and a displacement of 3780 tonnes. A weight of 12 tonnes is moved across the deck for a distance of 11 metres; when a plumb-line, suspended 8·0 metres above the batten, moves out 8 centimetres. Find the ship's light KG.

169. Find the KG of a ship which has a KM of 8·15 metres and displaces 2400 tonnes. When the inclining experiment was performed, a weight of 10 tonnes was shifted 16 metres across the deck and caused a plumb-line, 5 metres long, to move out 18 centimetres.

170. In an inclining experiment, a weight of 12·5 tonnes was moved 10 metres across the deck and caused a plumb-line, 12 metres long, to move out 32 centimetres. A double-bottom tank in the ship was full of water, which weighed 450 tonnes and had its centre of gravity 0·90 metres above the keel; otherwise the ship would have been in the light condition. If the ship's displacement at the time of the experiment was 3750 tonnes and her KM was 9·00 metres, find:—

(*a*) The KG at the time of the experiment.
(*b*) The light KG.

Answers—

166. 0·673 m.
167. 1·453 m.
168. 6·81 m.
169. 6·30 m.
170. (*a*) 8·00 m; (*b*) 8·97 m.

Moment of Statical Stability

171. Find the moment of statical stability of a ship of 3165 tonnes displacement and GM 0·80 metres, when she is heeled to an angle of 12°.

172. A ship of 1068 tonnes displacement has a GM of 1·20 metres. Find her moment of statical stability at an angle of heel of 6°.

173. What is the moment of statical stability of a ship which displaces 6752 tonnes and has a righting lever of 0·45 metres?

174. A ship of 5124 tonnes displacement has the following righting levers:—

Angle of heel:	10°	20°	30°	40°	50°	60°	70°
Corresponding GZ	0·12	0·33	0·48	0·52	0·39	0·18	—0·09 metres

Draw a curve of moments of statical stability and find from this:—

(*a*) The moment of statical stability at 24° of heel.

(*b*) The maximum moment of statical stability and the angle at which this occurs.

(*c*) The range of stability.

175. A ship of 7200 tonnes displacement has a *KB* of 4·00 metres and a *KG* of 6·50 metres. At an angle of heel of 23° the volumes of the immersed and emerged wedges are each 1200 cubic metres and the horizontal shift of their centres of gravity is 7·00 metres. Find the length of the righting lever and the moment of statical stability at this angle of heel.

176. A ship which is heeled to an angle of 57° has immersed and emerged wedges of 2500 cubic metres each, with their centres of gravity 12·0 metres apart. The ship displaces 12,500 cubic metres of salt water, has a *KG* of 7·10 metres and a *KB* of 4·30 metres. What is the moment of statical stability and is the ship in stable equilibrium?

177. A box-shaped ship is 120 metres long, 18 metres wide and floats at a draft of 4·00 metres. Assuming that the deck edge does not submerge, nor the bilge emerge, find the volumes of the wedges and the shift of their centres of gravity at an angle of heel of 25°. Thence find the righting lever at 25° of heel, if the *KG* is 8·30 metres.

Answers—

171. 526·7 t/m.
172. 134·6 t/m.
173. 3038·4 t/m.
174. 20,390 t/m; 37°; 26,900 t/m; 67°.
175. 0·218 m; 1569·6 t/m.
176. −0·03 m; Ship is unstable.
177. 0·66 m.

Angle of Heel

178. A weight of 50 tonnes is shifted transversely across the deck of a ship for a distance of 12 metres. The ship's displacement was 4350 tonnes and her *GM* was 0·40 metres. If the ship was upright before the weight was shifted, find the angle to which she will heel.

179. A ship of 4800 tonnes displacement has a list of 8°, due to unequal loading of weights. If her *GM* is 0·30 metres, find how much weight must be shifted transversely across a 'tween deck, for a distance of 12 metres, in order to bring the ship upright.

180. A weight of 120 tonnes is loaded into a 'tween deck so that it is 3·8 metres from the ship's centre-line and at a vertical height of 6·0 metres above her centre of gravity. Before the weight was loaded, the ship was upright, had a *GM* of 0·7 metres and a displacement of 7080 tonnes. What will be the effect of the weight?

181. A ship has a *KG* of 3·8 metres and displacement of 5750 tonnes and is listed 12° to port. 250 tonnes of cargo are to be loaded into the port and starboard wings of a 'tween deck at a height of 7·5 metres above the keel and a distance of 8·0 metres on either side of the centre-line. If the *KM* is 4·6 metres, how much weight must be placed in each wing to finish loading with the ship upright?

182. 80 tonnes of grain shifts in a hold, 10 metres horizontally and 3 metres downwards. Before this happened, the ship was upright, had a *GM* of 1·35 metres and a displacement of 2320 tonnes. Find the angle of heel caused by the shift of the grain.

183. A ship displaces 11600 t, has *KG* of 6·10 metres, *KM* of 6·95 metres and is heeled 5° to starboard. 350 t of cargo are to be loaded into the wings of a 'tween deck at distances of 5·0 metres to port and 7·0 metres to starboard of the centre line How much of the cargo must there be loaded into each wing in order to finish with the ship upright?

184. A ship displaces 14500 t, has *KG* of 7·10 metres, *KM* of 8·05 metres, and is heeled 4° to starboard. 500 t of cargo are to be loaded into a 'tween deck at a height of 10·0 metres above the keel: of this, 300 t are to be loaded in the square of the hatch, whilst the remainder is to be distributed between the wings, at 6·0 metres to port and 8·0 metres to starboard of the centre line. How much must be placed in each wing if the vessel is to be upright after it has been loaded?

185. A vessel is heeled 7° to starboard and has a *KG* of 6·02 metres, *KM* of 6·41 metres, and displaces 8800 t. She then loads 75 t cargo at 5·6 metres above the keel and 5·0 metres to starboard of the centre line; 100 t at 4·2 metres above the keel and 6·5 metres to port of the centre line; and 90 t amidships and at 3·8 metres above the keel. What will be her final heel, if any?

186. A ship displaces 12720 t, has *KG* of 6·900 metres, *KM* of 7·500 metres and is upright when a 60-tonne locomotive is stowed on deck, at 5·8 metres to starboard of the centre line and 11·5 metres above the keel. The locomotive is to be discharged over the port side of the ship by means of a derrick, the head of which will be 26 metres above the keel and 13 metres off the centre line when plumbing the quay. Find the maximum heel, the final heel and the final *GM*.

187. Two heavy lifts, each of 40 tonnes, are to be lifted from the quay by a derrick and are to be placed on the ship's deck at 12·0 metres above the keel and 6·0 metres on either side of the centre line. The first lift is to be landed on the offshore side of the deck and the second lift on the onshore side. The head of the derrick will be 22 metres above the keel and will plumb a point 11·5 metres from the centre line when lifting from the quay. Before lifting, the ship was upright, displaced 9600 t, had *KG* of 6·850 metres and *KM* of 7·460 metres. Find: (*a*) the maximum heel; (*b*) the heel when the first lift has been landed on deck and the second is being lifted from the quay.

188. A vessel displaces 12420 t and has a *KM* of 7·84 metres. She is to take on board a lift of 80 t, using a derrick, the head of which will be 12·5 metres off the centre line and 21 metres above the keel when lifting. What must be the ship's maximum *KG*, before lifting, in order that the heel may not exceed 5° when lifting?

189. Find the angle to which a ship will loll, if her *GM* is —0·06 metres and her *BM* is 3·60 metres.

190. A ship displaces 2040 tonnes and has a *GM* of 0·10 metres. 120 tonnes of cargo are then loaded on deck at a vertical height of 4·0 metres above the ship's centre of gravity, when the *BM* was found to be 3·20 metres. What will happen?

191. To what angle will a ship loll if she has a *BM* of 6·00 metres and a *GM* of —0·20 metres?

Answers—

178. 19°
179. 16·9 t.
180. 6·3° heel.
181. 64 t *(P)*; 186 t *(S)*.
182. 13°.
183. 278 t *(P)*; 72 t *(S)*.
184. 183 t *(P)*; 17 t *(S)*.
185. 2·1° to Starboard.
186. 9·5° *(P)*; 2·5° *(P)*; 0·622 m.
187. 5·0° *(S)*; 2·5° *(S)*.
188. 6·835 m.
189. 10·4°.
190. Ship lolls 15·3°.
191. 14·5°.

Free Surface Effect

192. A seaman calculates his ship's *GM* as 0·68 metres and her displacement as 4320 tonnes, when floating in salt water. He has forgotten to allow for free surface of salt water in a rectangular double-bottom tank, 15 metres long and 12 metres wide, with no subdivisions in it. What is the ship's true *GM* when upright?

193. If the tank in the last question had contained oil of relative density 0·875, what would then have been the *GM*?

194. A ship of 6000 tonnes displacement has a *KG* of 3·45 metres and a *KM* of 3·72 metres, neglecting the offect of free surface. Free surface exists in an undivided rectangular tank, 12 metres long and 10 metres wide, which is partly filled with sea water. Find the true *GM* when the ship is upright.

195. 100 tonnes of water are run into a rectangular tank, 10 metres long and 12 metres wide, in a ship of 5300 tonnes displacement: when the tank is found to be about three-quarters full. If the centre of gravity of this water is 0·50 metres above the keel and the ship's original *KG* was 5·00 metres, find the new *KG*.

196. A box-shaped lighter is 30 metres long and 8 metres wide and floats at a draft of 1·00 metre: whilst its *KG* is 0·80 metre. If 15 centimetres of water is then allowed to run into the bottom, so that the new *KM* becomes 5·20 metres, what will be the new *GM*?

197. A ship has a displacement of 4880 tonnes and a *KG* of 6·00 metres when all her double-bottom tanks are full of salt water. What would be the *KG* if 80 tonnes of water were pumped out of a rectangular tank, leaving it slack: the tank being 7·0 metres long and 16·0 metres wide, whilst the centre of gravity of the water removed was 5·00 metres below *G*.

198. A rectangular deep tank, 10 metres long, 12 metres wide and 5 metres deep is divided at the centre-line. When the tank is full, the ship has a displacement of 6080 tonnes and a *KG* of 6·00 metres. If the tank is then pumped out until there are 2·0 metres of water left in it, what will be the new *KG* of the ship, assuming that the tank extends right down to the keel?

199. A ship displaces 10400 t, has a *KG* of 6·45 metres, a *KM* of 7·52 metres and floats upright in salt water when the starboard side of a double bottom tank is full of salt water and the port side of the tank is empty. The tank is rectangular, 35 metres long, 16 metres wide and 1·6 metres deep; with one side girder on each side and a watertight centre girder. To what angle will the ship heel if exactly one half of the ballast is transferred from the starboard to the port side of the tank?

200. A rectangular tank is 24 metres long, 15 metres wide and has a fore and aft division at its centre line. What is its free surface moment?

201. Calculate the free surface moment of a rectangular double bottom tank which is 16 metres long, 18 metres wide, and has a centre girder and one side girder on each side.

202. A ship displaces 9850 t and has a solid *GM* of 1·080 metres when a tank is partly filled with oil of relatived ensity 0·910. If the tank has a free surface moment of 1220, what is the ship's fluid *GM*?

203. A vessel has a *KG* of 5·982 metres, a *KM* of 6·493 metres and displaces 7486 t. 200 tons of salt water are then run into a tank, leaving it slack. The centre of gravity of the water is then 0·65 metres above the keel and the tank has a free surface moment of 588. Find the ship's new fluid *GM*.

204. A ship displaced 10540 t, had a *KG* of 5·421 metres and a *KM* of 5·873 metres. She then discharged 150 t of salt water from 1·20 metres above the keel from a tank which had a free surface moment of 395: and 100 t of salt water from 1·30 metres above the keel from a tank which had a free surface moment of 457. Find the new fluid *GM*.

205. A ship loads 300 t of oil of relative density 0·895 at 0·85 metres above the keel in a tank which has a free surface moment of 512. She also loads 250 t of salt water at 0·70 metres above the keel in a tank which has a free surface moment of 386. Both tanks are then slack. The original displacement and KG were 8720 t, and 5·010 m respectively: whilst the final KM is 5·641 m. Find the new GM.

Answers—

192. 0·17 m.
193. 0·24 m.
194. 0·10 m.
195. 5·19 m.
196. −0·15 m.
197. 6·53 m.
198. 6·22 m.
199. 9·9°.
200. 1688.
201. 486.
202. 0·967 m.
203. 0·572 m.
204. 0·265 m.
205. 0·790 m.

Dynamical Stability

206. Find the dynamical stability at 40° heel of a ship which displaces 6876 tonnes and has the following GZs:—

Heel	10°	20°	30°	40°	
GZ	0·115	0·228	0·331	0·429	metres

207. Find the dynamical stability at 45° heel of a ship which has a displacement of 5000 tonnes and GZs of:—

Heel	15°	30°	45°	
GZ	0·205	0·482	0·611	metres

208. Find the area under a curve of righting levers up to 40° heel, given:—

Heel	10°	20°	30°	40°
GZ	0·145	0·248	0·264	0·234

Answers—

206. 1068 t/m.
207. 1312 t/m.
208. 0·138 metre-radians.

BM_L

209. A box-shaped ship has a length of 120 metres and floats at a draft of 8·00 metres. What is her longitudinal BM_L?

210. Find the BM_L and GM_L of a box-shaped lighter, 30 metres long, 10 metres wide and floating at a draft of 2·00 metres, if her KG is 1·50 metres.

211. A box-shaped lighter is 40 metres long and floats at a draft of 1·20 metres fore and aft. Find her BM_L and KM_L.

Answers—

209. 150 m.
210. 37·5 m; 37·0 m.
211. 111·1 m; 111·7 m.

Change of Draft due to Change of Trim

212. Find the new drafts if a weight is shifted aft in a ship, sufficiently to change the trim by 0·26 metres. The centre of flotation is amidships and the original drafts were 6·45 metres forward and 6·48 metres aft.

213. The centre of flotation of a ship is 4·0 metres abaft amidships. Weights are shifted aft so as to change the trim by 0·56 metres. If the ship's length is 140 metres and her original drafts were 5·92 m forward and 6·08 m aft, find the new drafts fore and aft and also the original and new mean drafts.

214. A ship is 120 metres long and floats on an even keel at drafts of 5·36 metres fore and aft. Weights are shifted aft so as to change the trim by 0·48 metres. Find the new drafts, fore, aft and mean, if the ship's centre of flotation is 5·0 metres abaft amidships.

215. An oil tanker is 260 metres long and floats at drafts of 5·12 m forward and 6·88 m aft. The centre of flotation is 4·0 metres forward of amidships. Oil is then shifted from an after tank to a forward tank so as to change the trim by 1·20 metres. Find the new drafts.

Answers—

212. F. 6·32 m; A. 6·61 m.
213. F. 5·62 m; A. 6·34 m; Old M, 6·00 m; New M, 5·98 m.
214. F. 5·10 m; A. 5·58 m; M. 5·34 m.
215. F. 5·70 m; A. 6·26 m.

Draft and Displacement Out of the Designed Trim

216. A ship which is 150 metres long, floats at drafts of 5·16 metres forward and 6·32 metres aft. If her designed trim is for an even keel and *F* is 4·0 metres abaft amidships, find the draft at *F*.

217. A vessel is 110 metres long, *F* is 3·0 metres forward of amidships and her designed trim is an even keel. If her drafts are 3·98 metres forward and 4·86 metres aft, what is the draft at *F*?

218. A ship is 140 metres long, has a T.P.C. of 20 and the centre of flotation is 1·5 metres abaft amidships. Find the layer correction when the drafts are 6·30 m forward and 6·70 m aft.

219. A ship is 133 metres long, has a T.P.C. of 19·0 and her centre of flotation is 3·0 metres abaft amidships. She floats at drafts of 6·12 m forward and 6·54 m aft. The displacement for the designed trim (even keel) at a draft of 6·33 m is 8243 tonnes. Find the layer correction and the true displacement.

220. A ship, 160 metres long, floats at drafts of 7·95 m forward and 8·59 m aft. The designed draft is an even keel: the T.P.C. is 25·0: and the centre of flotation is 2·5 metres abaft amidships. The displacement is given in the scale as 10,942 tonnes for a mean draft of 8·27 metres. Find the draft at *F* and the true displacement.

221. A ship is 90 metres long, her T.P.C. is 11·0 and her centre of flotation is 2.0 metres abaft amidships. The designed trim is 15 centimetres by the stern. The drafts are 4·15 m forward and 3·85 m aft. The designed displacement for a draft of 4·00 m is 3283 tonnes. Find the draft at *F* and the true displacement.

Answers—

216. 5·77 m.
217. 4·40 m.
218. 8·6 t.
219. 18 t; 8261 t.
220. 8·28 m; 10967 t.
221. 3·98 m; 3264 t.

M.C.T.1C.

222. A ship is 140 metres long and displaces 4340 tonnes, her KG is 5·10 metres and her KM_L 132·2 metres. Find her M.C.T.1C.

223. Find the M.C.T.1C. of a ship, 120 metres long and displacing 3600 tonnes, which has a GM_L of 150·0 metres.

224. A box-shaped vessel is 80 metres long, 15 metres wide and floats at a draft of 5·00 metres in sea water. KG is not known. Find her M.C.T.1C.

225. A ship is 192 metres long, has a KG of 6·80 metres, a KM of 200·2 metres and displaces 9200 tonnes. Find her M.C.T.1C.

Answers—

222. 39·4 t/m.
223. 45·0 t/m.
224. 82·0 t/m.
225. 92·7 t/m.

Change of Trim due to Shifting Weights

226. A ship floats at drafts of 6·84 m forward and 7·14 m aft. Her M.C.T.1C. is 105 tonne/metres and her centre of flotation is amidships. Find the change of trim and the new drafts if a weight of 42 tonnes is shifted forward for a distance of 60 metres.

227. Find the change of trim and the new drafts fore and aft, if 120 tonnes of oil is transferred from the forepeak to the afterpeak tank in a ship which has an M.C.T.1C. of 144 tonne/metres. The distance between the centres of gravity of the tanks is 60 metres, the ship's centre of flotation is amidships, and the original drafts were 4·82 metres forward and 4·58 metres aft.

228. What will be the change of trim and the new drafts fore and aft if a weight of 90 tonnes is shifted aft for a distance of 100 metres in a ship which has an M.C.T.1C. of 180? The length of the ship is 150 metres, her centre of flotation is 2·0 metres abaft amidships and her original drafts were 6·00 m forward and 6·10 m aft.

229. A ship is 160 metres long and her centre of flotation is 4 metres abaft amidships. Her drafts are 6·58 m forward and 8·08 m aft, whilst her M.C.T.1C. is 180 t/m. Find the change of trim and the new drafts if 300 tonnes of oil is transferred from No. 4 double-bottom tank to No. 1 double-bottom tank: a distance of 72 metres.

230. A ship has drafts of 4·10 m forward and 5·50 m aft. Her M.C.T.1C. is 200. Find how much oil to transfer from No. 5 D.B. tank to No. 2 D.B. tank, through a distance of 100 metres, to bring the ship to an even keel.

231. A ship has an M.C.T.1C. of 210 and floats at drafts of 6·20 m forward and 7·60 m aft. Find how much weight must be moved forward, through a distance of 60 metres, to bring her to a trim of 50 centimetres by the stern.

Answers—

226. 24 cm; F. 6·96 m; A. 7·02 m.
227. 50 cm; F. 4·57 m; A. 4·83 m.
228. 50 cm; F. 5·74 m; A. 6·34 m.
229. 120 cm; F. 7·21 m; A. 7·51 m.
230. 280 tonnes.
231. 315 tonnes.

Moderate Weights Loaded Off the Centre of Flotation

232. A ship is 150 metres long, has a T.P.C. of 12·5 and an M.C.T.1C. of 120. Her drafts are 4·76 m forward and 5·40 m aft. 250 tonnes of cargo are loaded at a distance of 50 metres abaft the stem. Find the new drafts, assuming that the centre of flotation is amidships.

233. The centre of flotation of a ship is amidships, her length is 120 metres, her T.P.C. is 15 and her M.C.T.1C. is 100. A weight of 240 tonnes is discharged from a position 40 metres abaft the stem. If her original drafts were 5·10 m forward and 5·20 m aft, what will be her new drafts?

234. The following particulars are known about a ship:—Length 136 metres; T.P.C. 20; M.C.T.1C. 120; centre of flotation 70 metres abaft the stem; drafts 6·50 m forward, 6·60 m aft. Find her new drafts after 80 tonnes of cargo have been loaded at a distance of 40 metres from aft.

235. A ship has drafts of 4·72 m forward and 5·64 m aft. In order to bring her more nearly to an even keel, 240 tonnes of water are run into No. 1 double bottom tank, the centre of gravity of which is 20 metres abaft the stem. The ship is 160 metres long, has a T.P.C. of 22, an M.C.T.1C. of 172 and the centre of flotation is 2·0 metres abaft the amidships. Find the new drafts.

236. Find the new drafts after 120 tonnes of cargo have been discharged from a point which is 15 metres abaft amidships. The ship is 130 metres long, has a T.P.C. of 15, and M.C.T.1C. of 110, whilst her centre of flotation is 3 metres abaft amidships. The original drafts were 7·00 m forward and 7·40 m aft.

237. A ship which is 130 metres long has a T.P.C. of 20 and M.C.T.1C. of 125. Her centre of flotation is amidships and her drafts are 6·20 m forward and 6·50 m aft. What would be the new drafts if 140 tonnes of cargo are loaded at a distance of 30 metres abaft the stem and 56 tonnes are loaded at a distance of 100 metres abaft the stem?

238. A ship floats at drafts of 4·30 m forward and 4·80 m aft. Her length is 108 metres, T.P.C. 16, M.C.T.1C. 118, whilst her centre of flotation is 2·0 metres abaft amidships. Find the new drafts after the following weights have been loaded and discharged:—

Loaded:	55 tonnes at 40 metres abaft the stem.
Loaded:	100 tonnes at 70 metres abaft the stem.
Discharged:	73 tonnes from 86 metres abaft the stem.

239. A vessel is 90 metres long and her centre of flotation is 2 metres abaft amidships. Her T.P.C. is 8·0 and her M.C.T.1C. is 48. 60 tonnes of cargo is discharged from 22 metres forward of the centre of flotation; whilst 40 tonnes is loaded at 13 metres abaft the centre of flotation. If the original drafts were 3·85 m forward and 3·79 m aft, find the new drafts.

240. A ship is 140 metres long and floats at drafts of 6·38 metres forward and 7·06 metres aft. Her T.P.C. is 22·1, M.C.T.1C. is 186 and F is 2·0 metres forward of amidships. She next takes in 310 t of water ballast at 6·0 metres abaft amidships and then sails for her next port. On the voyage, she uses 370 t of oil from 15 metres abaft amidships; 25 t of stores from 51 metres forward of amidships; and 48 t of fresh water from 3 metres abaft amidships. Find her draft on arrival.

Answers—

232. F. 5·22 m; A. 5·34 m.
233. F. 4·70 m; A. 5·28 m.
234. F. 6·45 m; A. 6·72 m.
235. F. 5·28 m; A. 5·33 m.
236. F. 6·99 m; A. 7·26 m.
237. F. 6·42 m; A. 6·48 m.
238. F. 4·42 m; A. 4·78 m.
239. F. 3·62 m; A. 3·94 m.
240. F. 6·39 m; A. 6·92 m.

Large Weights Loaded Off the Centre of Flotation

241. A ship floats at drafts of 4·90 metres forward and 5·10 metres aft. 600 tonnes of oil is then loaded into a deep tank, the centre of gravity of which is 3·0 metres forward of amidships. Find the new drafts, if the following information is found from the deadweight scales:—

Draft 5·0 m. T.P.C. 21·0. M.C.T.1C. 148
Draft 5·5 m. T.P.C. 21·8. M.C.T.1C. 154
Centre of flotation amidships.

242. The following information is given in the ship's deadweight scale:—

Draft 6·0 m. T.P.C. 16·1. M.C.T.1C. 108
Draft 6·5 m. T.P.C. 16·6. M.C.T.1C. 113
Draft 7·0 m. T.P.C. 17·0. M.C.T.1C. 117
Centre of flotation, 3·0 m abaft amidships

The ship is 126 metres long and floats at drafts of 6·10 m forward and aft. Find the new drafts if the following cargo is loaded:—

450 tonnes at 40 metres abaft the stem
500 tonnes at 110 metres abaft the stem

243. The ship, for which the hydrostatic particulars are given in the back of this book, is 140 metres long. She is floating at drafts of 5·16 metres forward and 5·24 metres aft. Use the curves or scales to find the new drafts after the ship has loaded:—

In No. 1 hold; 190 tonnes; c.g. 52 metres forward of amidships.
In No. 2 hold; 260 tonnes; c.g. 28 metres forward of amidships.
In No. 3 hold; 180 tonnes; c.g. 5 metres forward of amidships.
In No. 4 hold; 380 tonnes; c.g. 32 metres abaft amidships.
In No. 5 hold; 250 tonnes; c.g. 46 metres abaft amidships.

244. The same ship as in the last question had drafts of 2·58 metres forward and 4·72 metres aft. To bring her to a better trim, the forward deep tanks (c.g. 15·2 metres forward of amidships) were then filled with 1150 t of water. Find her new drafts.

245. The same ship, as above, when loading cargo, had drafts of 6·23 metres forward and 6·69 metres aft. She then loaded:—

480 tonnes at 48 metres forward of amidships.
710 tonnes at 24 metres forward of amidships.
630 tonnes at 32 metres abaft amidships.
370 tonnes at 43 metres abaft amidships.

This completed her loading and she then sailed for her next port. On the voyage she used 210 t of oil from 2 metres forward of amidships, 30 t of fresh water from 8 metres abaft amidships and 10 t of stores from 12 metres abaft amidships. What were her drafts on arrival?

Answers—

241. F. 5·24 m; A. 5·32 m.
242. F. 5·74 m; A. 7·88 m.
243. F. 5·58 m; A. 6·02 m.
244. F. 3·72 m; A. 4·71 m.
245. F. 7·25 m; A. 7·41 m.

Loading a Weight to Produce a Desired Trim

246. A ship, which is completing her loading, has 120 tonnes of cargo to come on board. Her drafts are 7·00 m forward and 7·82 m aft. Her M.C.T.1C. is 125 tonne-metres. Where must the cargo be loaded in order that the ship may sail with a trim of 50 centimetres by the stern?

247. Find the weight of water which must be run into a double-bottom tank in order to bring the ship on to an even keel. The centre of gravity of the tank is 25 metres abaft the stem. The ship is 140 metres long, has an M.C.T.1C. of 144 and her centre of flotation is 5·0 metres abaft amidships. Her present drafts are 5·70 m forward and 6·60 m aft.

248. A ship has been in collision and her fore peak is flooded, causing her to trim 0·60 m by the head. It is desired to bring her to a trim of 0·20 m by the stern. The after peak tank, which is empty, can take 240 tonnes of water and its centre of gravity is 70 metres abaft the ship's centre of flotation. If the M.C.T.1C. is 168, will it be possible to bring the ship to the desired trim by running up this tank and, if so, what weight of water must be taken in?

249. A ship floats on an even keel and has an M.C.T.1C. of 116. A total of 400 tonnes of cargo are to be loaded into No. 1 hold (60 m forward of F.) and into No. 4 hold (20 m abaft F.). How much cargo must be loaded into each hold in order that the ship may finish loading with a trim of 50 centimetres by the stern?

250. A ship floats at drafts of 5·60 m forward and 7·00 m aft. Her M.C.T.1C. is 140. 280 tonnes of water is then pumped out of No. 4 double-bottom tank, which has its centre of gravity at 15 metres abaft the centre of flotation. Calculate how much oil would have to be transferred from No. 6 double-bottom tank (55 m abaft F.) to No. 1 double-bottom tank (65 m forward of F.) to bring the vessel to an even keel.

Answers—

246. 33·3 m forward of F.
247. 259 tonnes.
248. Yes. 192 tonnes is required.
249. 27½ tonnes into No. 1;
372½ tonnes into No. 4.
250. 128 tonnes.

Loading for Required Draft Aft

251. A ship is 138 metres long, has an M.C.T.1C. of 132, a T.P.C. of 18, and her centre of flotation is 3 metres abaft amidships. How far forward of the centre of flotation must a weight be loaded if the after draft is to remain constant?

252. How far abaft the stem must a weight be loaded if the draft aft is not to change in a ship of 140 metres long? The centre of flotation is 2 metres abaft amidships, the T.P.C. is 22, and the M.C.T.1C. is 160.

253. The centre of flotation of a ship is amidships and her length is 120 metres. Her T.P.C. is 16 and her M.C.T.1C. is 115. Where, with relation to amidships, must a weight of 140 tonnes be loaded, if the draft aft is not to change?

254. A ship is 120 metres long and has drafts of 5·83 metres forward and 5·49 metres aft. Her T.P.C. is 16·4, M.C.T.1C. is 115 and *F* is amidships. How much cargo must she load into an after hold, the centre of gravity of which is 48 metres abaft amidships, in order to increase the draft, aft, to 6·00 metres?

255. An oil tanker, when light, floats at drafts of 2·62 metres forward and 5·24 metres aft. Her length is 200 metres, T.P.C. is 38, M.C.T.1C. is 190, whilst *F* is 5·0 metres forward of amidships. On sailing, she has to cross a bar, on which the depth is 5·0 metres, with a clearance of 50 cm. Find the least amount of water ballast to load into a trimming tank, the centre of gravity of which is 95 metres forward of amidships. Find, also, the final draft forward.

256. A ship is 140 metres long and has drafts of 7·54 metres forward and 7·68 metres aft. Her T.P.C. is 23·0, M.C.T.1C. is 207, whilst *F* is 2·0 metres abaft amidships. Her fore peak tank is full of water ballast, with its centre of gravity 64 metres forward of amidships. Calculate the amount of this water ballast to be discharged in order to bring the ship to a draft, aft, of 8·00 metres. What would then be the draft forward?

257. A ship has 400 tonnes of cargo to load and her present drafts are 6·16 metres forward and 6·92 metres aft. She is 126 metres long, her T.P.C. is 20·0, M.C.T.1C. is 158, and *F* is 1·0 metres abaft amidships. How far from amidships should the cargo be loaded in order to bring the after draft to 7·00 metres?

Answers—

251. 15·3 m.
252. 57 m.
253. 14·4 m forward.
254. 189 t.
255. 333 t; 3·46 m.
256. 287 t; 6·94 m.
257. 8·6 m forward.

Weight to Load for a Given Draft

258. A ship has drafts of 5·86 metres forward and 6·12 metres aft. Her M.C.T.1C. is 114 and *F* is 2·0 metres abaft amidships. She has 220 tonnes of cargo to load into two holds, one at 56 metres forward of amidships and the other at 19 metres abaft amidships. How much should be loaded into each hold in order to bring the ship to a trim of 10 cm. by the stern?

259. A ship arrives off a port with drafts of 7·12 metres forward and 7·88 metres aft. Her T.P.C. is 19·5, M.C.T.1C. is 110 and her centre of flotation is amidships. In order to enter the port, she has to reduce her draft to not more than 7·40 metres. Find the minimum amount of cargo which she must discharge into lighters from two holds, one 32 metres forward of amidships and the other 54 metres abaft amidships.

260. A vessel has a T.P.C. of 16·3, an M.C.T.1C. of 115, whilst *F* is 2·0 metres abaft amidships. Her present drafts are 6·18 metres forward and 6·40 metres aft. In order to cross a bar, her maximum draft on sailing must not exceed 6·50 metres. Find the maximum amount of cargo which she can load into each of two holds, one 44 metres forward of amidships and the other 35 metres abaft amidships.

261. A ship which is loading cargo has to load 240 tonnes into No. 3 hold, at 12 metres forward of amidships; then to distribute as much cargo as possible between No. 1 hold (49 metres forward of amidships) and No. 4 hold (32 metres abaft amidships). On sailing, she has to cross a bar, on which the depth is 6·00 metres, with a clearance of 20 cm. What is the maximum amount of cargo to be loaded into each hatch, if her present drafts are 5·37 metres forward and 5·61 metres aft? The ship's T.P.C. is 20·1, her M.C.T.1C. is 166, whilst *F* is amidships.

262. In a Tropical Zone, a ship which has a summer draft of 7·94 metres, arrives in port with salt water drafts of 7·98 metres forward and 8·16 metres aft. Her T.P.C. is 23·2, M.C.T.1C. is 208, *F* is 3·0 metres abaft amidships and her Fresh Water Allowance is 164 mm. She then has to cross a dock sill, where the relative density of the water is 1·010 and where her mean draft must not exceed 8·00 metres. How much cargo must she discharge, before entering the dock, from each of two holds, one of which is 18 metres forward of amidships, and the other 36 metres abaft amidships?

Answers—

258. Forward, 74 t; Aft, 146 t.
259. Forward, 25 t; Aft, 170 t.
260. Forward, 175 t; Aft, 167 t.
261. No. 1, 165 t; No. 4, 218 t.
262. Forward, 171 t; Aft, 223 t.

The Use of Moments About the After Perpendicular

263. A ship displaces 9870 t and *B* is 58·25 metres from the after perpendicular. She loads 750 t at 22·0 metres from the after perpendicular. If *B* is then 58·37 metres from the after perpendicular, what is the moment changing trim?

264. A vessel displaces 5260 t and her *B* is 59·72 metres from the A/P. She then loads 250 t at 94 metres from the A/P and 320 t at 35 metres from the A/P: she also discharges 180 t from 16 metres from the A/P. If *B* is then 59·74 metres from the A/P, find the moment changing trim.

265. The following is an extract from the ship's hydrostatic information:—

Draft	Displacement	*B* from the A/P
6·00 m.	8147 t.	58·558 m.
6·20 m.	8461 t.	58·516 m.
6·40 m.	8792 t.	58·477 m.

Her present drafts are 6·08 metres forward and 6·04 metres aft. She then loads 260 t at 84 metres from the A/P, 430 t at 37 metres from the A/P, and discharges 180 t from 54 metres from the A/P. What is the moment changing trim?

266. A vessel is 120 metres long and floats at drafts of 5·28 metres forward and 6·14 metres aft. At this draft she displaces 7620 t, her T.P.C. is 15·94, *B* is 58·76 metres from the A/P, whilst *F* is 58·02 metres from the A/P.

She then loads 920 t of cargo at 67·80 metres from the A/P. The T.P.C. is then 16·18, M.C.T.1C. is 114·0, *B* is 58·67 metres from the A/P and *F* is 57·90 metres from the A/P. Find the new drafts.

267. A ship is 164 metres long and has drafts of 8·02 metres forward and 8·10 metres aft. At this draft her displacement is 18050 t, T.P.C. is 27·80, *B* is 82·14 metres from the A/P, whilst *F* is 78·94 metres from the A/P. She then:—

Loads 810 t at 69 metres from the A/P.
Discharges 650 t from 118 metres from the A/P.
Discharges 430 t from 64 metres from the A/P.
Discharges 720 t from 56 metres from the A/P.

At the new drafts, the T.P.C. is 27·76, M.C.T.1C. is 248, *B* is 82·32 metres from the A/P and *F* is 78·99 metres from the A/P. Find the new drafts.

268. The hydrostatic information for a ship which is 122 metres long, shews:—

Draft	Displacement	M.C.T.1C.	*B* from A/P	*F* from A/P
6·00 m.	8128 t	112·4	60·091 m.	59·258 m.
6·20 m.	8409 t	113·6	60·068 m.	59·169 m.
6·40 m.	8702 t	114·7	60·045 m.	59·079 m.
6·60 m.	9007 t	115·7	60·021 m.	58·988 m.

Her present drafts are 5·47 metres forward and 6·59 metres aft. She then loads 520 t at 113 metres from the A/P, 330 t at 98 metres from the A/P, and 410 t at 17 metres from the A/P. She also discharges 490 t from 85 metres from the A/P. What are her new drafts?

269. A vessel which is 150 metres long, floats at drafts of 4·28 metres forward and 4·24 metres aft. At this draft she displaces 7520 t and the centre of buoyancy is 77·30 metres forward of the after perpendicular. She then loads:—

No. 1 hold: 290 t at 114 metres from the A/P.
No. 2 hold: 670 t at 95 metres from the A/P.
No. 3 hold: 930 t at 82 metres from the A/P.
No. 4 hold: 780 t at 41 metres from the A/P.
No. 5 hold: 510 t at 24 metres from the A/P.
Oil fuel: 400 t at 85 metres from the A/P.

At the ship's new displacement, the draft at *F* is 6·01 metres. The M.C.T.1C. is 194·2 t/m, *B* is 76·43 metres from the A/P, whilst *F* is 74·56 metres from the A/P. Find her new drafts.

270. The ship for which the hydrostatic particulars are given in the back of this book is 140 metres long and floats at drafts of 2·74 metres forward and 3·60 metres aft. She then loads 1100 t of water ballast into a deep tank, at 77·2 metres from the A/P, and sails on a voyage. During the voyage she used 380 t of fuel from 80·1 metres from the A/P, 100 t of fresh water from 50·5 metres from the A/P, and 20 t of stores from 71·4 metres from the A/P. What were her drafts on arrival at her next port?

Answers—

263. 28461 t/m by the stern.
264. 8416 t/m by the head.
265. 1277 t/m by the stern.
266. F. 6·26 m; A. 6·32 m.
267. F. 7·44 m; A. 7·94 m.
268. F. 6·48 m; A. 6·66 m.
269. F. 5·62 m; A. 6·40 m.
270. F. 3·23 m; A. 3·75 m.

Hydrostatic Curves and Scales

The curves and scales given in the back of this book are for a vessel of 140 metres long; light draft 2·967 metres; summer load draft 8·094 metres; and light *KG* of 6·623 metres. Use them, as appropriate, to solve the following questions.

271. Extract all possible information from (i) the hydrostatic particulars; (ii) the deadweight scale; (iii) the hydrostatic curves, for each of the following drafts in salt water: (*a*) 4·20 metres; (*b*) 5·50 metres; (*c*) 6·34 metres; (*d*) 7·49 metres.

272. Find the displacement and deadweight at the summer draft.

273. What is the ship's light *GM*?

274. Find the ship's fresh water allowance.

275. The ship is loading in a dock, where the relative density of the water is 1·012, and her present drafts are 8·04 metres forward and 8·12 metres aft. How much more cargo can she load in order to be at her summer load line in salt water if she expects to use 42 t of fuel, water and stores on the way from the dock to the open sea?

276. The vessel is loading in a tropical zone, in water of relative density 1·015. The lower edge of the tropical load line is 74 mm. above water on the port side, whilst the upper edge of the summer load line is level with the water on the starboard side. How much more cargo can she load in order to be at her tropical load line in salt water?

277. The ship anchors off a port with drafts of 8·12 metres fore and aft, in water of density 1·026 t/m³. On the way to her berth, she has to cross a dock still on which the depth of water is 8·20 metres and the density is 1·018 t/m³. Find the least amount of cargo which the ship must discharge into lighters, before proceeding to her berth, in order to cross the dock sill with a clearance of 15 cm., assuming that she remains on an even keel throughout.

278. The ship has a *KG* of 6·61 metres and drafts of 7·13 metres forward and 7·85 metres aft. She then loads 720 t at 6·2 metres above the keel; 615 t at 9·5 metres above the keel; and 430 t of deck cargo at 11·4 metres above the keel. Find the new *GM*.

279. The vessel is empty of cargo and has only the following weights on board:—

Fuel	630 t at	0·6 metres above the keel.
Fresh water	95 t at	0·7 metres above the keel.
Stores	60 t at	10·0 metres above the keel.

She then completely fills a deep tank with 1102 t of water ballast at 4·84 metres above the keel, leaving no free surface effect. Find her *GM*s before and after filling the tank. How do you account for the change in *GM*?

280. The ship, which has a deck cargo of 660 t of timber at 12·0 metres above the keel, commences a voyage with drafts of 6·72 metres forward and 6·88 metres aft and *KG* of 7·67 metres. During the voyage she uses 480 t of fuel from 0·6 metres above the keel, 50 t of fresh water from 0·7 metres above the keel, and 20 t of stores from 7·0 metres above the keel; whilst the timber increases its weight by 15% through absorption of water. Find her *GM* on arrival and the angle of loll.

281. The ship sails with drafts of 6·83 metres forward and 6·91 metres aft. *KG* of 6·77 metres, and all fuel and fresh water tanks full (no free surface effect). During the voyage she expects to use:—

8 t of stores from 10·4 metres above the keel.

30 t of fresh water from 9·3 metres above the keel, leaving the tank slack (Free surface moment 104 t/m).

210 t of oil fuel from 9·3 metres above the keel, leaving the tank empty.

300 t of oil fuel from 0·8 metres above the keel, leaving the tank slack (free surface moment 378 t/m and density of oil 0·950 t/m³).

Estimate the ship's *GM* on arrival at her next port.

282. When a deep tank is 100% full of water of relative density 1·020, the ship has drafts of 4·04 metres forward and 4·20 metres aft, whilst her *GM* is 3·08 metres. 920 t of water, with its centre of gravity at 5·47 metres above the keel, is then pumped out of the tank, leaving it slack. If the free surface moment of the tank is 1278 t/m, find the ship's new *GM*.

283. The ship has drafts of 7·18 metres forward and 7·50 metres aft. It is desired to load 200 t of cargo, so as to maintain the after draft at 7·50 metres. Where should the cargo be placed and what will be the final draft forward?

284. The vessel is loading cargo and has drafts of 6·72 metres forward and 6·88 metres aft. She is to load cargo into No. 1 hold (115 metres from the A/P) and No. 4 hold (40 metres from the A/P). How much cargo must be loaded into each hold in order to finish with the ship on an even keel with drafts of 7·00 metres fore and aft?

285. The ship has drafts of 4·93 metres forward and 5·57 metres aft, whilst her *KG* is 7·28 metres. She then loads:—

480 t at 123 metres from the A/P and at 8·8 metres above the keel.
720 t at 101 metres from the A/P and at 5·7 metres above the keel.
560 t at 78 metres from the A/P and at 6·3 metres above the keel.
610 t at 36 metres from the A/P and at 7·1 metres above the keel.
370 t at 23 metres from the A/P and at 8·5 metres above the keel.

She then sails and on voyage uses the following fuel, water and stores:—

425 t from 80 metres from the A/P and 0·7 metres above the keel.
56 t from 121 metres from the A/P and 0·8 metres above the keel.
19 t from 71 metres from the A/P and 9·0 metres above the keel.

Find her drafts and solid *GM* on arrival.

Answers—

271.

	(*a*)	(*b*)	(*c*)	(*d*)
Displacement	6893	9534	11331	13890
Deadweight	2334	4975	6772	9331
T.P.C.	19·64	20·99	21·75	22·74
M.C.T.1C.	140·1	164·2	180·1	199·5
LCB from A/P	72·33	71·85	71·50	70·97
LCF from A/P	71·17	70·06	69·20	68·02
KB	2·27	2·99	3·45	4·09
KM	8·66	8·16	8·03	7·99
KM_L	286·9	244·2	226·2	205·1

272. 15279 t; 10720 t.
273. 3·777 m.
274. 164 mm.
275. 269 t.
276. 461 t.
277. 284 t.
278. 1·19 m.
279. 3·75 m; 3·18 m.
280. −0·02; 5½°.
281. 0·99 m.
282. 3·32 m.
283. 86·08 m from A/P; 7·36 m.
284. No. 1, 210 t; No. 4, 236 t.
285. F. 6·19 m, A. 6·41 m; 0·53 m.

Stability Curves

286. Use the cross curves given in Fig. 77 to find the righting levers for:—

(*a*) Heel 30°, Displacement 9000 t, *KG* 7·00 metres.
(*b*) Heel 60°, Displacement 6700 t, *KG* 7·50 metres.
(*c*) Heel 45°, Displacement 7800 t, *KG* 6·75 metres.
(*d*) Heel 15°, Displacement 5600 t, *KG* 7·20 metres.

287. From the cross curves in Fig. 77, find the righting levers for a *KG* of 7·00 metres and displacement of 5000 t. Use these to draw a curve of statical stability and from this find the range of stability, amount and angle of maximum stability and the approximate *GM*.

288. Use the cross curves (Fig. 77) to find the *Gz*s for a displacement of 8500 t and *KG* of 6·50 metres. Draw a curve of statical stability and find the angle of vanishing stability and the approximate *GM*.

289. Use the *KN* curves given in Fig. 78 to find the righting levers for:—

(*a*) Heel 30°, Displacement 6000 t, *KG* 6·00 metres.
(*b*) Heel 45°, Displacement 7900 t, *KG* 7·25 metres.
(*c*) Heel 15°, Displacement 9800 t, *KG* 6·43 metres.
(*d*) Heel 75°, Displacement 8500 t, *KG* 6·92 metres.

290. Use the *KN* curves given in Fig. 78 to find the righting levers for s displacement of 7600 t and a *KG* of 6·78 metres. From these, draw a curve of statical stability and find the range of stability and the amount and angle of maximum stability.

291. A ship has a *KG* of 6·56 metres and a *KM* of 6·45 metres. Her *KN*s are as follows:—

Heel	5°	10°	15°	30°	45°	60°
KN (m)	0·56	1·13	1·72	3·43	4·80	5·63

Find the righting levers, draw a curve of statical stability and find the range of stability, the approximate angle of loll and the amount and angle of maximum stability.

Answers—

286. (*a*) 0·37 m; (*b*) 0·18 m; (*c*) 0·79 m; (*d*) 0·26 m.
287. 85°; 1·02 m at 44°; 1·02 m.
288. 85°; 0·77 m.
289. (*a*) 1·17 m; (*b*) 0·41 m; (*c*) 0·28 m; (*d*) 0·05 m.
290. 85°; 0·79 m at 45°.
291. 57°; 12; 0·19 m at 38°.

The Metacentric Diagram

292. Construct a metacentric diagram for drafts of between 3·00 and 8·00 metres for the ship for which the hydrostatic information is given in the back of this book. From this, find *KB, KM* and *BM* for drafts of (*a*) 4·50 metres and (*b*) 6·25 metres.

293. Calculate the *KM* and *BM* of a box-shaped lighter, 30 metres long and 6 metres beam, for every half-metre of draft from 1·00 to 4·00 metres. Construct a metacentric diagram and from this find the *KB*, *KM* and *BM* for drafts of (*a*) 2·60 metres and (*b*) 3·40 metres.

Answers—

292.	(*a*)	*KB* 2·44 m	*KM* 8·50 m	*BM* 6·06 m.
	(*b*)	*KB* 3·40 m	*KM* 8·04 m	*BM* 4·64 m.
293.	(*a*)	*KB* 1·30 m	*KM* 2·46 m	*BM* 1·16 m.
	(*b*)	*KB* 1·70 m	*KM* 2·59 m	*BM* 0·89 m.

Bilging

294. Find the permeability of the following cargoes:—

(*a*) Stowage factor 2·60; Relative density 1·12
(*b*) Stowage factor 0·40; Relative density 8·00.
(*c*) Stowage factor 1·50; Relative density 1·75.

295. A box-shaped lighter, 30 metres long and 8 metres wide, floats at drafts of 1·00 metres fore and aft. It is divided into three equal compartments by two transverse bulkheads. Find the new drafts if the centre compartment, which is empty, is holed below the waterline.

296. What would have been the draft, in the last question, if the compartment had been filled with cargo of permeability 40%?

297. A box-shaped vessel is 75 metres long, 12 metres beam and floats at a draft of 6·20 metres fore and aft. A compartment amidships is 15 metres long and has a permeability of 60%. Find the new drafts if this compartment is bilged.

298. A box-shaped vessel is 60 metres long, 12 metres beam and 5·75 metres deep. She floats on an even keel at a draft of 4·80 metres. What will happen if an empty compartment amidships, 12 metres long, is bilged?

299. A ship is 120 metres long, 18 metres beam and floats at a mean draft of 6·00 metres. The coefficient of fineness of the waterplane is 0·750. A rectangular compartment amidships is 15 metres long, extends for the full width and depth of the ship, and has a permeability of 60%. Find the sinkage if this compartment is bilged.

300. A box-shaped lighter, 30 metres long and 8 metres beam, floats at drafts of 1·20 metres fore and aft. Find the sinkage and the new drafts if an empty compartment, right forward and 3·0 metres long, is bilged.

301. A box-shaped vessel is 80 metres long, 15 metres beam and floats at drafts of 3·00 metres fore and aft. Find the sinkage and new drafts if an empty compartment, right forward and 12 metres long, is bilged.

302. A ship, 120 metres long, floats at drafts of 4·50 metres fore and aft. The waterplane area is 1400 square metres, the displacement is 5800 tonnes, M.C.T.1C. is 96, whilst B is 1·50 metres abaft amidships. At this draft, the forepeak, which is empty, has a volume of 50 cubic metres below water, a waterplane area of 25 square metres, whilst its centre of gravity is 3·5 metres abaft the stem. Find the sinkage and change of trim if the forepeak is bilged.

303. A box shape, 110 metres long and 12 metres beam, floats on an even keel at a draft of 5·00 metres. An empty compartment, 10 metres long, has its centre of gravity 30 metres forward of amidships. Find the new drafts if this compartment is bilged. M.C.T.1C. is 102·4 tonne-metres.

304. A box-shaped vessel, 72 metres long and 7 metres beam, floats at drafts of 4·00 metres fore and aft. An empty compartment, right forward, is 6 metres long and has a watertight flat 3·0 metres above the keel. Find the new drafts if this compartment is bilged below the flat.

305. A box-shaped lighter is 30 metres long and 7 metres beam. She floats at a draft of 1·50 metres fore and aft. An end compartment, 5 metres long, has a watertight flat 1·50 metres above the keel and has a permeability of 45%. Find the new drafts if this compartment is bilged.

306. A ship is 120 metres long and floats on an even keel at a draft of 6·00 metres. Her displacement is 8000 tonnes; M.C.T.1C. is 110; the waterplane area is 1680 square metres; *B* and *F* are both 2·0 metres abaft amidships. The after peak has a capacity of 240 tonnes of salt water and its centre of gravity is 53 metres abaft *F*. Find the new drafts if the after peak is bilged.

Answers—

294. (*a*) 66%; (*b*) 68%; (*c*) 62%.
295. 1·50 m.
296. 1·15 m.
297. 6·74 m.
298. Vessel sinks.
299. 0·67 m.
300. 0·13 m; F. 1·87 m; A. 0·88 m.
301. 0·53 m; F. 6·06 m; A. 1·66 m.
302. 3·6 cm; 31 cm.
303. F. 6·45 m; A. 4·65 m.
304. F. 4·93 m; A. 3·57 m.
305. F. 1·89 m; A. 1·33 m.
306. F. 5·54 m; A. 6·70 m.

Drydocking

307. A ship enters a drydock with drafts of 3·00 metres forward and 3·50 metres aft. Her displacement is 3000 tonnes. *KG*, 7·30; *GM*, 2·00; M.C.T.1C., 88. The centre of flotation is 55 metres from aft. What will be the ship's *GM* at the instant of settling on the blocks, fore and aft?

308. In the case of the ship in the last question, what would be her *GM* when she was flat on the blocks and the water-level had fallen to 2·80 metres, her displacement then being 2500 tonnes?

309. The ship is trimmed 60 centimetres by the stern when she enters a dock. Her displacement is 4300 tonnes; *KG*, 6·90 m; *KM*, 8·40 m; M.C.T.1C., 100: whilst the centre of flotation is 70 metres from aft. Find the *GM* at the instant before the ship comes flat on the blocks fore and aft.

310. A box-shaped vessel is 100 metres long, 12 metres beam and floats at drafts of 2·40 metres fore and aft. Her *KG* is 4·90 metres. Find her new *GM* when she is flat on the blocks in a drydock and the water-level has fallen so that the draft is 2·00 metres fore and aft.

Answers—

307. 1·75 m.
308. 0·45 m.
309. 1·33 m.
310. 0·27 m.

INDEX

INDEX

HYDROSTATIC CURVES & SCALES

HYDROSTATIC PARTICULARS (in Salt Water)

DRAFT	DIS-PLACE-MENT	T.P.C.I. Tonnes Per Cm. Immer-sion	M.C.T.C. Moment to Change Trim One Cm.	L.C.B. Fwd. of A.P.	L.C.F. Fwd. of A.P.	V.C.B. A.B.	K.M. (T.) Trans-verse Meta-centre A.B.	K.M. (L.) Longl. Meta-centre A.B.
Metres	*Tonnes*			*Metres*	*Metres*	*Metres*	*Metres*	*Metres*
2·40	3550	17·38	109·3	72·79	72·50	1·29	11·62	427·4
·60	3901	17·68	112·6	72·76	72·36	1·40	11·16	402·8
·80	4257	17·96	116·0	72·72	72·22	1·51	10·73	382·1
3·00	4619	18·21	119·4	72·68	72·07	1·62	10·33	363·5
·20	4986	18·47	122·8	72·63	71·93	1·73	9·94	346·9
·40	5358	18·72	126·2	72·57	71·78	1·84	9·58	332·8
·60	5735	18·97	129·6	72·51	71·63	1·94	9·26	319·3
·80	6117	19·20	133·1	72·45	71·48	2·05	9·01	306·9
4·00	6502	19·42	136·6	72·39	71·33	2·16	8·82	296·3
·20	6893	19·64	140·1	72·33	71·17	2·27	8·66	286·9
·40	7296	19·86	143·7	72·26	71·01	2·38	8·55	278·2
·60	7696	20·07	147·3	72·19	70·85	2·49	8·45	270·3
·80	8107	20·28	150·9	72·12	70·68	2·60	8·36	263·5
5·00	8496	20·50	154·6	72·05	70·51	2·71	8·29	257·5
·20	8908	20·69	158·4	71·97	70·33	2·82	8·23	251·9
·40	9323	20·89	162·2	71·89	70·15	2·93	8·18	246·6
·60	9744	21·09	166·1	71·81	69·96	3·04	8·13	241·7
·80	10167	21·27	170·0	71·73	69·76	3·15	8·09	237·1
6·00	10596	21·44	173·8	71·65	69·56	3·26	8·06	232·9
·20	11027	21·63	177·5	71·56	69·35	3·37	8·04	228·9
·40	11461	21·80	181·2	71·48	69·14	3·49	8·02	225·0
·60	11899	21·98	184·8	71·39	68·92	3·60	8·00	221·2
·80	12340	22·15	188·3	71·29	68·71	3·71	7·99	217·4
7·00	12786	22·32	191·7	71·20	68·50	3·82	7·99	213·7
·20	13234	22·49	195·0	71·11	68·30	3·93	7·99	210·1
·40	13686	22·66	198·1	71·01	68·11	4·04	7·99	206·6
·60	14140	22·83	201·2	70·91	67·92	4·15	8·00	203·2
·80	14599	23·00	204·2	70·82	67·73	4·26	8·01	200·0
8·00	15060	23·17	207·1	70·72	67·54	4·37	8·02	196·9
·20	15525	23·34	210·0	70·62	67·36	4·48	8·03	194·1
·40	15994	23·51	212·8	70·52	67·19	4·59	8·05	191·2
·60	16467	23·68	215·5	70·41	67·03	4·71	8·07	188·4

(A.P. = After Perpendicular) (A.B. = Above Base Line)

HYDROST

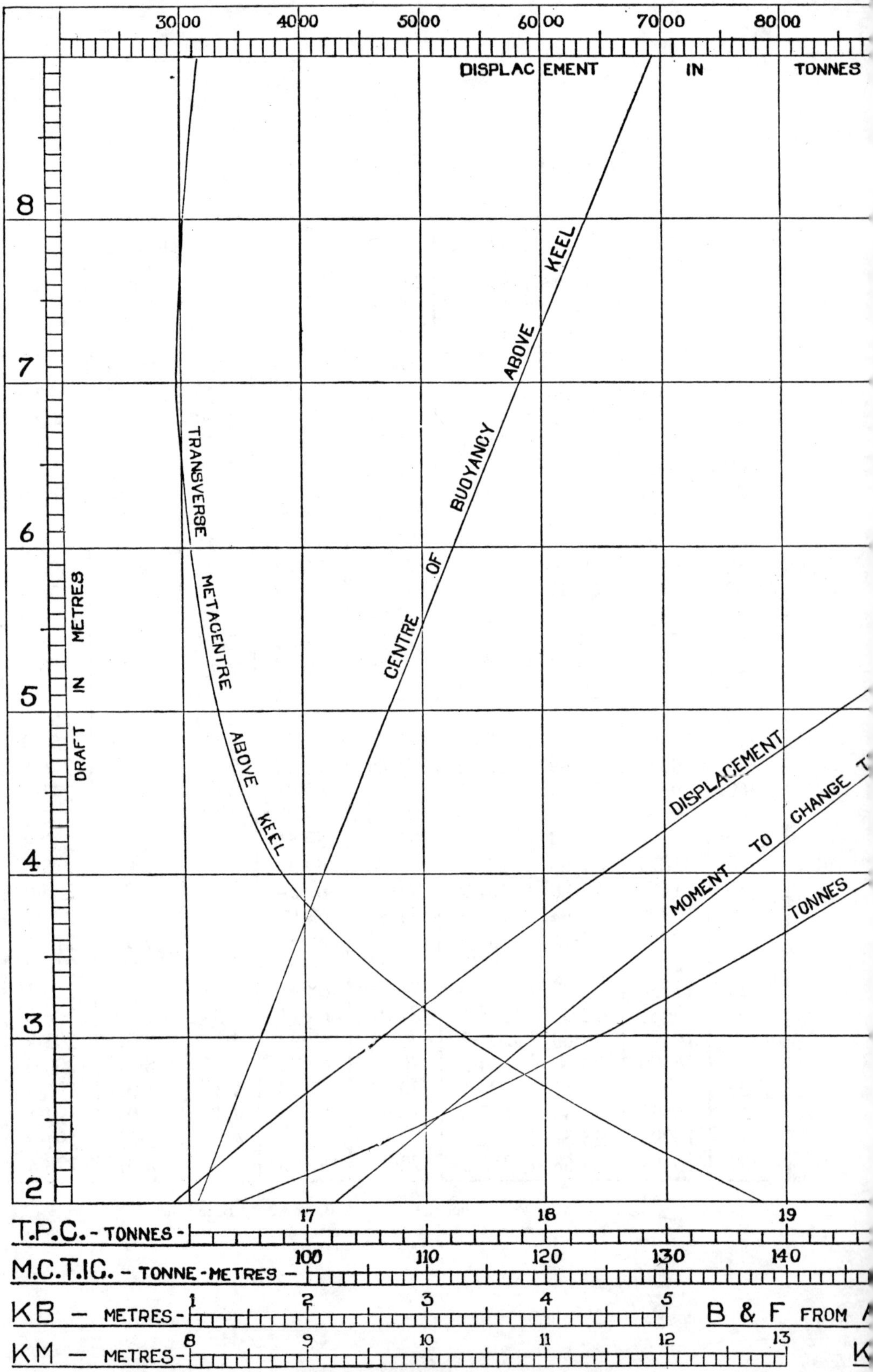

Inserted in *Merchant Ship Stability*. By H. J. Pursey. Published by Brown, Son & Ferguson, Ltd., Glasgow G41 2SD *Copyright*.

CURVES

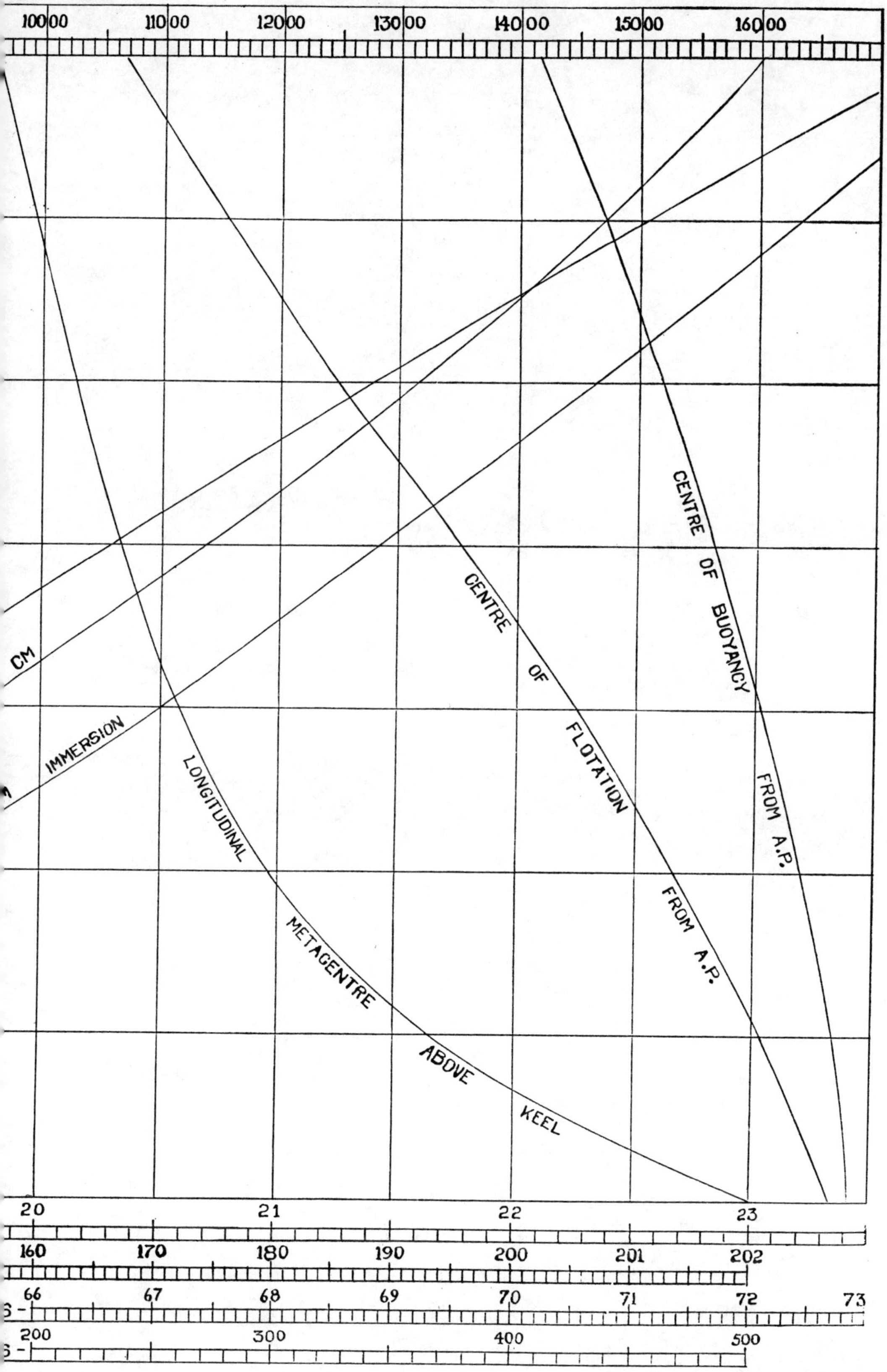

10000
11000
12000
13000
14000
15000
16000
CENTRE OF BUOYANCY FROM A.P.
CENTRE OF FLOTATION FROM A.P.
LONGITUDINAL METAGENTRE ABOVE KEEL
IMMERSION
CM
20
21
22
23
160
170
180
190
200
201
202
66
67
68
69
70
71
72
73
200
300
400
500

DEADWEIGHT SCALE

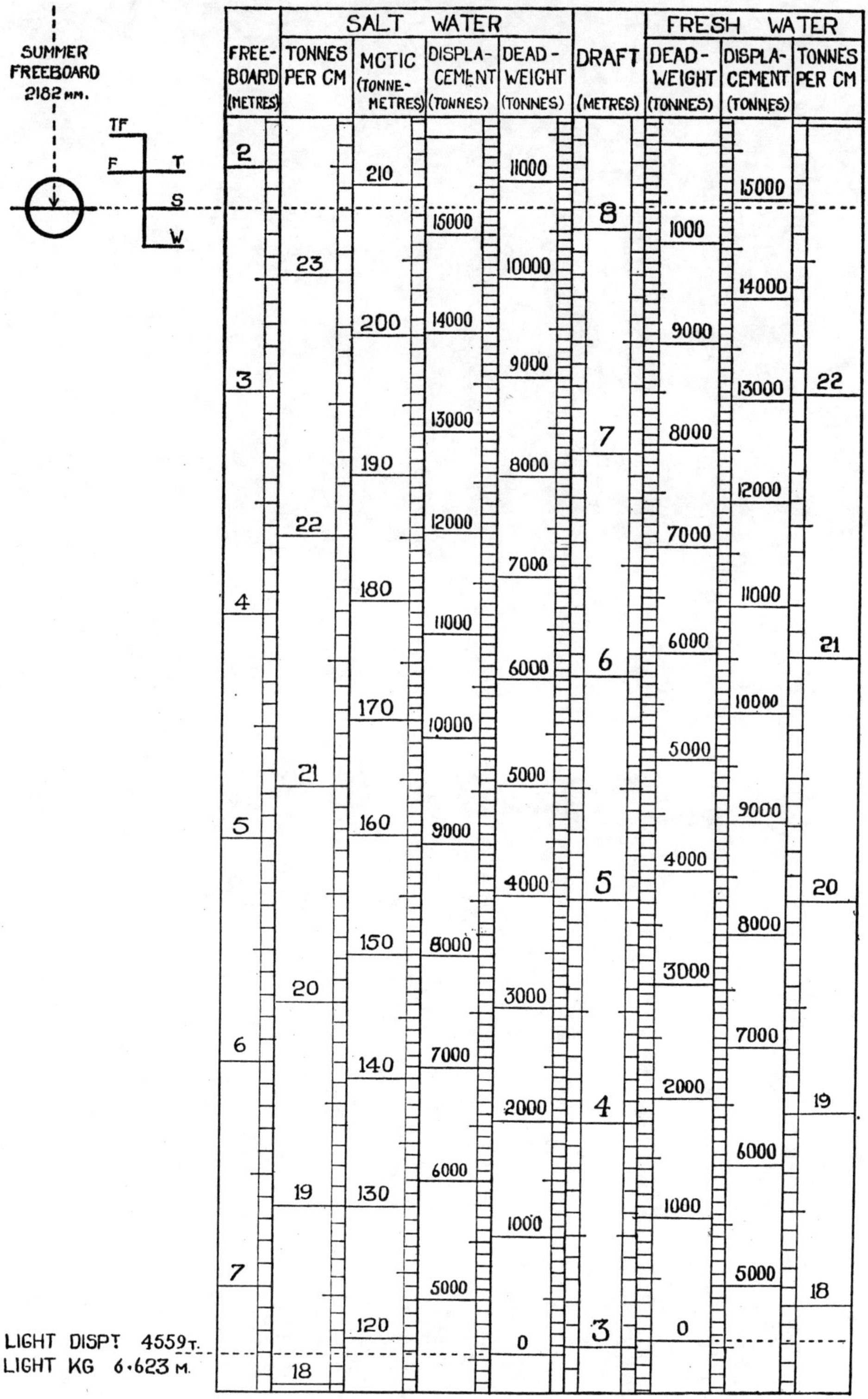